Phase Change Materials - Technology and Applications

Edited by Manish Rathod

Published in London, United Kingdom

Phase Change Materials - Technology and Applications
http://dx.doi.org/10.5772/intechopen.100743
Edited by Manish Rathod

Contributors
Sunny Chandra, Om Prakash, Anupam Prakash, Ricardo Scheepers, Ronald Coutu, Farhana Anwar, Turja Nandy, Mahboob E. Afshan, Anna Gowsalya Lucas, Abdelwaheb Trigui, Pilar Ocón, Alberto F. Romero, Yang Yang, Yuanfeng Xu, Desheng Fu, Yiming Zhang, Hao Zhang, Yujie Xia, Juan Zhang, Manish K. Rathod

First published in London, United Kingdom, 2022 by IntechOpen
IntechOpen is the global imprint of INTECHOPEN LIMITED, registered in England and Wales, registration number: 11086078, 5 Princes Gate Court, London, SW7 2QJ, United Kingdom

British Library Cataloguing-in-Publication Data
A catalogue record for this book is available from the British Library

Additional hard and PDF copies can be obtained from orders@intechopen.com

Phase Change Materials - Technology and Applications
Edited by Manish Rathod
p. cm.
Print ISBN 978-1-80356-473-9
Online ISBN 978-1-80356-474-6
eBook (PDF) ISBN 978-1-80356-475-3

Meet the editor

Dr. Manish K. Rathod is an assistant professor in the Department of Mechanical Engineering, at the National Institute of Technology (SVNIT), Surat, India. He received a Ph.D. from SVNIT in 2014. He has published more than seventy journal articles and has four design patents to his credit. He has presented numerous research papers at international conferences. In 2020 and 2021, he was ranked among the top 2% of world scientists in the subject of energy by Stanford University, USA. His areas of interest are thermal energy storage using phase change material (PCMs), thermal management by PCMs, passive cooling in buildings, energy and exergy analysis of thermal systems, and design and augmentation of heat exchangers.

Contents

Preface

The development of thermal appliances and equipment that make use of non-conventional energy sources is currently receiving a significant amount of focus within the realm of thermal energy research. The ability to store thermal energy is a fundamental prerequisite for such applications. When it comes to thermal energy storage (TES), the use of phase change material (PCM) for latent heat storage is more desirable than sensible heat storage. This is due to its potential to store a large amount of energy for a given volume at a reduced temperature swing. Other reasons include its compactness, uniform energy storage, and delivery, and temperature swings.

This book provides readers with a comprehensive overview of PCMs and their current applications. It begins with a discussion of the basic concept of phase-change energy storage and the categorization of phase-change energy materials. Next, the book explores the essential features that influence the use of PCMs, followed by an in-depth discussion of the techniques for the thermal analysis of PCM. Other topics addressed include recent uses of PCMs in electronic cooling, battery thermal management, and textile and solar energy conversion.

Section 1 provides a brief introduction to PCMs and the different techniques used to enhance their thermal performance. Section 2 discusses the design of a heat exchanger that is based on latent heat as well as methods for conducting thermal analysis of PCMs. Section 3 discusses the application of PCMs in battery thermal management. Section 4 explores the use of PCMs to cool electronic devices and Section 5 discusses the applications of PCMs in the textile industry. Section 6 focuses on other recent applications of PCMs.

I would like to express my thanks to IntechOpen for the care and attention it has shown in publishing this book on schedule.

Manish Rathod
Department of Mechanical Engineering,
Sardar Vallabhbhai National Institute of Technology,
Surat, India

Section 1

Introduction to Phase Change Material

Chapter 1

Introductory Chapter: Phase Change Material as Energy Storage Substance

Manish K. Rathod

1. Introduction

The economic success and technical competitiveness of a country are significantly influenced by its access to energy. World energy outlook [1] reported that the total energy demand for the countries like India and China is expected to increase from 55% in 2010 to 65% in 2035. Due to factors such as population growth, fast urbanisation, and economic development, demand for easily accessible energy sources like fossil fuels is rising at an almost exponential rate. Despite the development of resources and the expansion of energy supplies, enough energy has not been produced to fulfil the rising demand. Additionally, the demand for energy in various sectors like residential and commercial, industrial, and utility fluctuates on a time basis. In order to meet such variation, it is essential to look for suitable means of storage of energy.

The storage of energy offers an alternate approach to meeting peak demand for electricity. The most expensive and challenging times to provide are during peak hours. Therefore, energy storage systems offer a tremendous amount of untapped potential to improve the efficiency of equipment that converts energy. It also has the potential to be utilised for enabling large-scale fuel replacements in economies all around the world. Energy storage results in cutting down the energy wastage. This is a cheaper, easier, and faster way to address several energy issues. In a nutshell, energy storage not only acts as a link between demand and availability but also boosts the system's overall performance and dependability. Consequently, the development of energy storage systems that are efficient and low-cost is an equally vital subject of study as the development of new sources of energy.

2. Energy storage systems

The act of temporarily 'storing' energy for the purpose of using it at a later time is referred to as energy storage (ES). Power and energy services rely on ES devices, which take in excess energy and hold it for later use. Energy may be stored in a variety of forms, including mechanical, electrical, chemical, and thermal energy. The ES systems are generally categorised by output: electricity and thermal.

2.1 Electricity storage systems

Electricity storage systems include hydropower, compressed air energy, flywheels, batteries, supercapacitors, chemical-hydrogen storage, etc.

- Pumped Storage Hydropower (PSH) systems take use of fluctuations in elevation to accumulate off-peak electricity for use at a later time. During times of lesser demand, water is moved by pump from one reservoir to another reservoir located at a higher elevation. After that, the water is released to begin its descent back to the lower reservoir, which results in the generation of electricity in a manner analogous to that of a simple hydropower plant.

- Compressed Air Energy Storage (CAES) systems utilise energy generated during non-peak hours in order to compress air. This compressed air is then stored in underground storage tanks. Later on, this air is sent out into the combustor of a gas turbine, which helps create energy during peak demand times.

- Flywheels that spin at very high rates can be used to store electricity in the form of rotational energy. The rotor of the flywheel is slowed down to release this stored energy in small bursts.

- Batteries facilitate the flow of electrons by using chemical processes that take place inside two or more electrochemical cells. Sodium sulphur batteries, lead-acid batteries, and lithium-based batteries are some examples.

- Energy is stored in supercapacitors by enormous electrostatic fields that are created between two conducting plates that are very slightly separated from one another. This technique allows for the rapid storage and release of electrical energy, which may result in the production of brief surges of power.

- Chemical hydrogen storage is a method of storing electricity where hydrogen is used as the energy carrier. After going through a series of transformations, electrical energy is stored until it is needed.

2.2 Thermal storage systems

The requirement of a thermal storage system arises in many applications. These include harnessing intermittent sources of energy like solar, wind, etc. and heat recovery systems. Thus, developing efficient and reliable thermal storage devices is an important requirement for both conventional and non-conventional thermal energy systems. Thermal storage system includes sensible heat storage (i.e. solid media storage, hot- and cold-water storage, underground thermal storage), latent heat storage (i.e., ice storage, molten salts), and thermochemical storage.

2.2.1 Sensible heat storage

Practical Heat Storage (SHS) is the process of storing heat by increasing the surrounding temperature. Increasing the temperature of a substance, whether it be a solid or a liquid, serves as a means of storing thermal energy in this system. In SHS systems, the amount of stored heat depends on the storage material, the specific heat of the substance, and the change in temperature. The need for a high volume of material is one of the most significant drawbacks of SHS, particularly when the acceptable temperature range is narrow.

2.2.2 Latent heat storage

Latent heat storage units (LHSUs) are used to store energy in a substance that changes phase when heat is added or removed. When a medium goes from one state to another—solid, liquid, or gas—it is said to have undergone a phase change. Whether energy is being absorbed or discharged determines the direction of this transition (i.e., liquid to solid or solid to liquid).

In comparison to SHS systems, LHSU is a more appealing option. This is due to its high energy density and very narrow temperature range. LHSU also offers consistent energy storage and delivery in a compact design. In the case of water, for instance, the latent heat of fusion is eighty times that of the sensible heat. As a result, at a constant temperature of 0 degrees Celsius, latent heat-based storage systems that utilise water may extract 80 times as much energy as a sensible heat-based storage system. This implies that the quantity of material for a specific amount of energy may be drastically reduced in both weight and size. It is to be noted that the energy storage capacity of a LHSU is generally 5–10 times higher than that of an SHS system [2–5].

Among the several energy storage technologies discussed above, some technologies have undergone significant research and demonstrations and have matured for commercialization. These include pumped storage technologies, sensible heating technologies etc. Some technologies like thermochemical, supercapacitor, flywheel (high speed), flow batteries, etc., are still in the preliminary stage of research. Technologies like latent heat storage units (LHSU), flywheel (low speed), compressed air energy storage, etc., however fall in the category where there need more research and demonstrations before they can be brought to commercialization. More work has to be done on these systems before they can reach their full potential.

3. Phase change materials

In an LHSU, energy is kept in the latent heat storage material. The term "Phase Change Materials" refers to the types of substances that may store latent heat (PCMs). Telkes and Raymond [6] were among the first to pioneer the research of PCMs. However, before the energy crises of the 1970s and 1980s, nobody paid any attention to it, when there was intensive research on the use of PCMs in many applications, particularly for solar heating systems. There have been several PCMs found and investigated extensively for their application in LHSU during the last four decades, and these PCMs span a wide range of melting/freezing points. A classification of these PCMs is given in **Figure 1**.

Many factors, including thermodynamic, chemical, kinetic, and economic ones, go into deciding which PCM to use. To be considered as PCMs, a substance must meet the following criteria:

- When choosing a PCM for the first time, it's important to make sure its melting point falls within the system's optimal working temperature range.

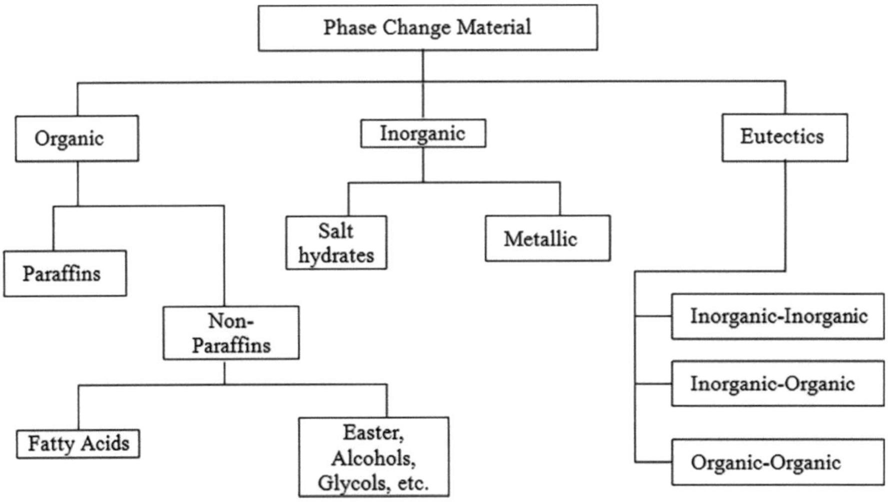

Figure 1.
Classification of PCMs [5].

- To alleviate the containment issue, the PCM should have a low vapour pressure at an operational temperature, a modest volume expansion during the phase transition, and a high latent heat.

- Insignificant super cooling as the super cooling will reduce the heat storage capacity of PCMs.

- High thermal conductivity to enhance the high heat transfer rates and also high specific heat.

- A high nucleation rate is also required for kinetic properties to reduce supercooling and a high rate of crystal formation.

- Chemicals with non-toxic, non-explosive, and non-flammable properties are included in the category of chemical qualities. The capacity to completely reverse the melting/solidification cycle, chemical stability, the absence of any chemical breakdown, and corrosion resistance to building materials are all important properties.

- Abundant supply at minimum cost.

At least three components are needed for an LHSU: a PCM with an appropriate melting point, a heat exchange surface, and a PCM-compatible container. Thus, Phase Change Materials (PCMs), Containers, and Heat Exchangers must be understood to build an efficient LHSU.

Paraffin and nonparaffin organics are organic substances. Organic compounds offer noncorrosiveness, less cooling, chemical stability and thermal stability, harmonious melting ability, and suitability with common construction materials. Because of the broad temperature ranges across which it may melt or solidify and its large latent heat capacity, paraffin is a frequently used PCM for heat storage. During the process of solidification, they do not cause any subcooling effects and experience just a little volume change during the phase transition. They are stable, nontoxic, and noncorrosive over time. Lauric, myristic, palmitic, and stearic acids are used as PCMs in non-paraffin organics. Predicting melting and solidification behaviour and eliminating subcooling effects are their advantages. Organic compounds are combustible, have poor heat conductivity, and low phasechange enthalpy. Inorganic substances include salt hydrates, metals, alloys, and salts. Salt hydrates were explored, including sodium sulphate decahydrate (Glauber's salt), calcium chloride hexahydrate, etc. Inorganic compounds have excellent thermal conductivity and volumetric latent heat storage, sometimes double that of organic compounds. Most salt hydrates have problems with super cooling, phase segregation, corrosion, and thermal stability.

3.1 Drawbacks of PCM

PCM with lower thermal conductivity has a major effect on the efficiency of the device. Relatively large temperature reductions are observed during the energy withdrawal or retrieval process when conductivity values are reduced. Because of this, PCM melting and solidification rate has been slower than predicted, and the deployment of large-scale LHTS units has proved unsuccessful. Incomplete melting/solidification and a wide temperature disparity inside the PCM are common outcomes of this scenario, and they can contribute to the eventual failure of the material and overheating of the system. Therefore, it is even more crucial to develop certain thermal improvement measures to improve the thermal performance of LHSU.

3.2 Thermal enhancement techniques

The most studied thermal performance enhancement techniques are as follows: -

1. Use of extended surface/Fins

2. Micro Encapsulation

3. High thermal conductivity Nano particles added to the PCM

4. Add PCM into the Porous metallic foams

3.2.1 Use of extended surface/fins

Fins implanted in the PCM are the most widely used of all of these approaches. Fin configurations in the PCM-LHTES are classed as longitudinal, circular/annular, plate, and annular/pin fin. The selection was influenced by the benefits of a larger heat-transfer surface, simplicity, ease of production, and cheap cost of construction. Fins are utilised in thermal systems to provide greater heat transmission surface. Various academics have looked into the significance of different fin layouts in improving performance in the LHSU.

3.2.2 Micro encapsulation

Microencapsulated PCMs (MPCMs) may also improve heat transmission between the source and PCM. Microencapsulated PCMs are micro-sized PCMs that are either liquid or solid at the core and are surrounded by a solid shell or wall. There is a vast range of possible materials that may be used to construct the shell, including synthetic and natural polymers. Such MPCMs can be accomplished using a variety of chemical (e.g., coacervation, complex coacervation, interfacial methods), mechanical, or physical processes (e.g. spray drying method). It is to be anticipated that the thermal performance of MPCMs will outperform that of PCMs that are used traditionally. The heat transmission rate is greater for small PCM particles due to larger heat transfer area per unit volume. There are also advantages to using MPCMs, such as the ability to withstand changes in volume during a phase transition and less PCM reactivity with the container material.

3.2.3 High thermal conductivity nano particles added to the PCM

In spite of the fact that the impact is heavily dependent on the dispersion of the nanomaterial additions in PCM, carbon nanomaterial additives can enhance the PCM's thermal conductivity. During this process, the nanomaterial is distributed evenly throughout the material, resulting in uniform composite phase change material (CPCM). For the purpose of ensuring that the effectively enhanced thermal conductivity while simultaneously avoiding chemical reactions, the additives need to have a high thermal conductivity and be chemically stable. Metal oxides, carbon nanomaterials such as single- and multi-walled carbon nanotubes, metal nanoparticles, graphite and graphene are commonly used as additives to increase the thermal conductivity of PCM.

3.2.4 Add PCM into the porous metallic foams

Metal matrices constructed of aluminium, copper, and other metals, as well as naturally occurring porous materials like graphite, can be used to create porous structures. Incorporating high thermal conducting material with a PCM storage system improves heat transfer (latent heat phase). The capillary forces and surface tension in the combination of PCM and ceramic structure keep the molten PCM contained and stabilised inside the micro-porosity of the structure, enabling the use of direct contact heat exchangers. Graphite's higher thermal conductivity, lower density, and chemical resistance make it a promising heat transfer enhancer. The fraction of the composite that is composed of expanded graphite has a significant impact on the effective thermal conductivity of the material, as seen in **Figure 2**. There should be limits placed on the graphite content since increasing it will drive up production costs and reduce the material's volume storage capacity.

3.2.4.1 Closure

Due to the fact that no one material exists with all the desirable characteristics of an ideal thermal storage media, it is necessary to make do with the material at hand and, if necessary, compensate for its subpar physical characteristics through innovative system design. Thus, selecting a PCM is challenging for researchers. In addition to its usage in the energy storage system, PCMs have a wide range of other potential uses in a variety of fields. A number of researchers have reportedly completed a review on the technical particulars of various PCMs and the applications for which they are used. **Table 1** is a summary of some of the more important uses of PCMs that have been cited by a variety of researchers as having undergone rigorous review.

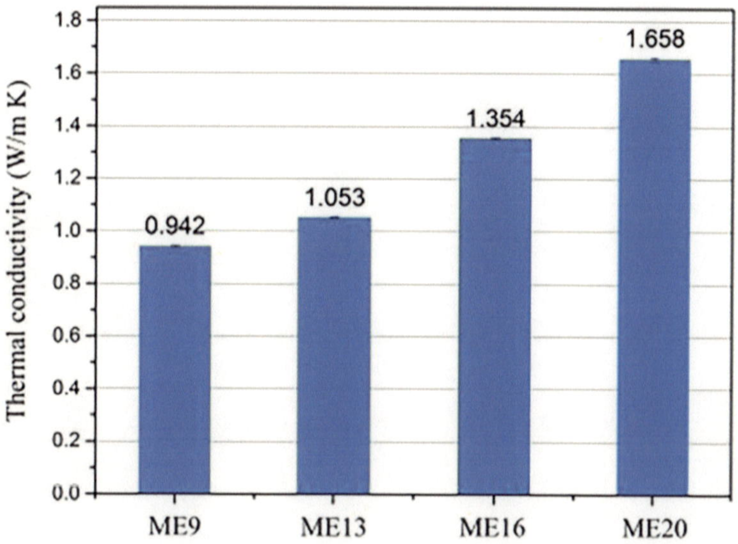

Figure 2.
Obtained thermal conductivity of PCMs with $MgCl_2 \cdot 6H_2O$/EG for different mass fractions of expanded graphite (EG) [7].

Applications	References
Latent heat storage for solar heating systems	[8–10]
Waste heat recovery	[11–14]
Building air conditioning	[15–18]
Building envelope	[19–21]
Refrigeration system	[22, 23]
Electronics cooling	[24–26]
Textiles	[27, 28]
Foods	[29, 30]

Table 1.
Applications of phase change materials (PCMs).

Inadequate LHSU performance is caused by the poor heat conductivity of commercially available PCMs, variation in thermo-physical properties after number of thermal cycles, phase separation, volume expansion and high cost [5]. There is a need to devise mechanisms for augmentation in thermal performance of LHSU under these circumstances. Thus, for development of state-of-the-art LHSU, comprehensive investigations are required incorporating the effect of all operating, geometric and design parameters of interest. The results of these analyses can be used as a starting point for further process optimization and the development of design principles.

Author details

Manish K. Rathod
Department of Mechanical Engineering, Sardar Vallabhbhai National Institute of Technology, Surat, India

*Address all correspondence to: mkr@med.svnit.ac.in

References

[1] World Energy Outlook. International Energy Agency. 2017. ISBN 978-92-64-28230-8

[2] Dincer I, Rosen MA. Thermal Energy Storage Systems and Applications. 2nd ed. United Kingdom: John Wiley and Sons Ltd; 2002

[3] Farid MM, Khudhair AM, Siddique AK, Sari A. A review on phase change energy storage: Materials and applications. Energy Conversion and Management. 2004;**45**:1597-1615

[4] Kaygusuz K. The viability of thermal energy storage. Energy Sources. 1999;**21**:745-755

[5] Rathod MK, Banerjee J. Thermal stability of phase change materials used in latent heat energy storage systems: A review. Renewable and Sustainable Energy Reviews. 2013;**18**:46-258

[6] Telkes M, Raymond E. Storing solar heat in chemicals—A report on the Dover house. Heating and Venting. 1949;**46**(11):80-86

[7] Song Z, Deng Y, Li J, Nian H. Expanded graphite for thermal conductivity and reliability enhancement and supercooling decrease of MgCl2·6H2O phase change material. Materials Research Bulletin. 2018;**102**:203-208

[8] Shukla R, Sumathy K, Erickson P, Gong J. Recent advances in the solar water heating systems: A review. Renewable and Sustainable Energy Reviews. 2013;**19**:173-190

[9] Guelpa E, Verda V. Thermal energy storage in district heating and cooling systems: A review. Applied Energy. 2019;**252**:113474

[10] Ray AK, Rakshit D, Ravikumar K. High-temperature latent thermal storage system for solar power: Materials, concepts, and challenges. Cleaner Engineering and Technology. 2021;**4**:100155

[11] Yagi J, Akiyama T. Storage of thermal energy for effective use of waste heat from industries. Journal of Materials Processing Technology. 1995;**48**:793-804

[12] Vasiliev LL, Burak VS, Kulakov AG, Mishkinis DA, Bohan PV. Latent heat storage modules for preheating internal combustion engines: Application to a bus petrol engine. Applied Thermal Engineering. 2000;**20**(10):913-923

[13] Nomura T, Okinaka N, Akiyama T. Waste heat transportation system, using phase change material (PCM) from steelworks to chemical plant. Resources, Conservation and Recycling. 2010;**54**:1000-1006

[14] Omara AAM. Phase change materials for waste heat recovery in internal combustion engines: A review. Journal of Energy Storage. 2021;**44**:103421

[15] Kuznik F, David D, Johannes K, Roux J. A review on phase change materials integrated in building walls. Renewable and Sustainable Energy Reviews. 2011;**15**(1):379-391

[16] Zalba B, Marn JM, Cabeza LF, Mehling H. Free-cooling of buildings with phase change materials. International Journal of Refrigeration. 2004;**27**:839-849

[17] Pasupathy A, Velraja R, Seeniraj RV. Phase change material-based building architecture for thermal management in residential and commercial

estab- lishments. Renewable and Sustainable Energy Reviews. 2008;**12**:39-64

[18] Al-Yasiri Q, Szabo M, Arici M. A review on solar-powered cooling and air-conditioning systems for building applications. Energy Reports. 2022;**8**:2888-2907

[19] Scalat S, Banu D, Hawes D, Parish J, Haghighata F, Feldman D. Full scale thermal testing of latent heat storage in wallboard. Solar Energy Materials and Solar Cells. 1996;**44**:49-61

[20] Zhou G, Yang Y, Wang X, Zhou S. Numerical analysis of effect of shape-stabilized phase change material plates in a building combined with night ventilation. Applied Energy. 2009;**86**:52-59

[21] Bhamare DK, Rathod MK, Banerjee J. Passive cooling techniques for building and their applicability in different climatic zones—The state of art. Energy & Building. 2019;**198**:467-490

[22] Liu M, Saman W, Bruno F. Validation of a mathematical model for encapsulated phase change material flat slabs for cooling applications. Applied Thermal Engineering. 2011;**31**:2340-2347

[23] Raut D, Kalamkar VR. A review on latent heat energy storage for solar thermal water-lithium bromide vapor absorption refrigeration system. Journal of Energy Storage. 2022;**55**:105828

[24] Lu TJ. Thermal management of high power electronics with phase change cooling. International Journal of Heat and Mass Transfer. 2000;**43**:2245-2256

[25] Tan FL, Tso CP. Cooling of mobile electronic devices using phase change materials. Applied Thermal Engineering. 2004;**24**:159-169

[26] Hua W, Zhang L, Zhang X. Research on passive cooling of electronic chips based on PCM: A review. Journal of Molecular Liquids. 2021;**340**:117183

[27] Ying BA, Kwok YL, Li Y, Zhu QY, Yeung CY. Assessing the performance of textiles incorporating phase change materials. Polymer Testing. 2004;**23**:541-549

[28] Sarier N, Onder E. Organic phase change materials and their textile applications: An overview. Thermochimica Acta. 2012;**540**:7-60

[29] Gin B, Farid MM. The use of PCM panels to improve storage condition of frozen food. Journal of Food Engineering. 2010;**100**:372-376

[30] Martin V, He B, Setterwall F. Direct contact PCM–water cold storage. Applied Energy. 2010;**87**:2652-2659

Section 2

Phase Change Material Technology

Chapter 2

Technology in Design of Heat Exchangers for Thermal Energy Storage

Mahboob E. Afshan and Anna Gowsalya Lucas

Abstract

In today's world, the energy requirement has full attention in the development of any country for which it requires an effective and sustainable potential to meet the country's needs. Thermal energy storage has a complete advantage to satisfy the future requirement of energy. Heat exchangers exchange heat in the thermal storage which is stored and retrieved later or can be used as a pre-heating or post-heating devices to save energy. Criteria of design of heat exchangers for various thermal energy storage applications along with their various components are being elaborated. The latent heat thermal energy storage in a mass application has got many advantages over the sensible heat storage. The existing approaches in the design, integration and application of phase change materials (PCMs) are explored by experimenting on a prototype of a single heat exchanger module and analysing all the design aspects to get a significant idea and the analysis needed while charging and discharging the heat exchanger.

Keywords: thermal energy storage, heat exchanger, design, latent heat, PCM

1. Introduction

Thermal energy storage (TES) systems can be employed for both heating and cooling applications. TES is a process of storing heat from various sources like waste heat or solar thermal applications or electricity used at off-peak rates or can also be used in cooling applications. The heat transfer fluid (HTF) at low temperature is stored and used in peak hours of heating TES. The estimated market by the research study of the global thermal energy storage market was 4,281.6 Million USD in 2019 and is anticipated to grow upto USD 8558.34 Million by 2026. An annual compound rate of growth of the thermal energy storage market is going to be 10.4% from 2019 to 2027 as reported.

Various types of new and innovative TES are available and used for various purposes as follows:

1. Cooking and Greenhouse

2. Building passive heating and cooling from stable PCMs

3. Performance enhancement techniques

4. Steam accumulators in solar parabolic trough

5. Power generating fields

6. Desalination and still water

7. Production/crop drying

The TES can be combined with waste heat recovery or solar thermal heating or cooling applications. In heating, electric-resistance efficiencies (near 100% on an energy basis) combined with lower off-peak electric rates can produce heating at a fraction of the cost of conventional systems. One of the applications in cold countries is to produce heat at night with air heating using electrical resistance heater and stored in such storage media like ceramic bricks in insulated containers. The heat produced due to electric off-peak rates in the night is advantageous and generally 33–75% cheaper than the peak rates. On demand of the space heating, the stored heat is transferred to the room in the peak hours.

Another important application is the Concentrating Solar Power (CSP) plants which are important solar energy technologies where sunlight is converted to high-temperature heat by concentrating it in mirrors and lenses from 25 to 3000 times of the solar light intensity. If heat transfer fluids like sodium are used, heat at high temperatures (3000^0C) can be retrieved. Sunlight being an intermittent source of energy that can produce heat helps to fill the gap between the energy demand and supply. A thermal energy storage is designed based on the diurnal, seasonal or annual storage needs and integrated into the system to sustain full or partial load operation.

This thermal energy can be stored at any particular temperature by using sensible or latent heat or both sensible and latent heat and can also be designed based on the application which can always be used in tropical countries due to the availability of solar radiation. A line diagram of a CSP with a two-tank TES taken from Ref. [1] is depicted in **Figure 1** where the line-concentrating solar panels concentrate the light and generate high-quality heat, which heats the heat transfer fluid. The heat transfer fluid in turn heats the incoming steam in the super heater. This two-tank molten salt TES system used in CSP is a technology commercially employed.

1.1 Objectives of thermal storage

The main objective of thermal storage is to deal with the amount of heat being stored at a particular temperature based on the application. The disparity of energy when supplied and consumed along with the necessity to store the surplus energy which would or otherwise go to waste including the shifting peak energy or power demand suggests the use of thermal energy storage for various areas of application such as space heating, hot water and air conditioning. Thermal energy storage facilitates superior and efficient use of the sources of fluctuating energies by harmonising the supply and demand of energy.

The methods used for thermal energy storage apply for less than 8 hours in northern winters, and the economical supply of the required energy becomes limited when compared to the costs of the combined solar collector and thermal energy storage system in these winters. An alternate heating source is taken as a backup for

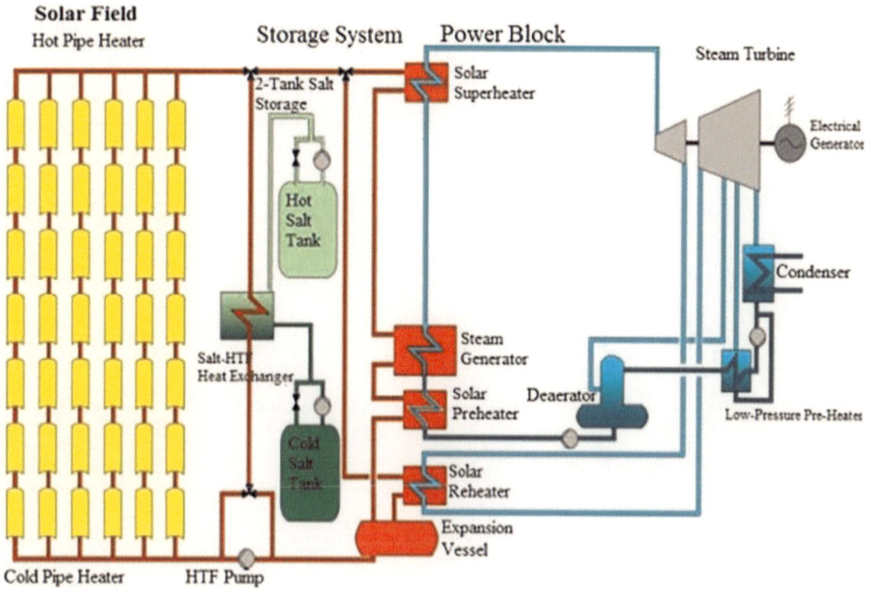

Figure 1.
Line diagram for a two-tank thermal energy storage CSP system.

extended sunless periods but it limits [2] the use of solar energy further. However, an economically acceptable storage system will take advantage of low off-peak electricity prices, which enhances the benefits in thermal energy storage.

1.2 Types of thermal energy storage

Three types of TES systems are categorised as sensible heat thermal energy storage, latent heat thermal energy storage and thermochemical energy storage [1, 3].

1.2.1 Sensible thermal energy storage

Storing of sensible energy due to the virtue of increase or decrease of temperature for a storage material is called sensible thermal energy storage. Air, water, rock, brick and concrete are a few sensible heat storage materials. Based on each material's advantages and disadvantages, it is preferred subject to the specific heat capacity including volume occupied by the TES [3]. To quantify sensible thermal energy stored, the following equation is given as follows:

$$Q = m.Cp.\Delta T \quad (1)$$

Q represents the total quantity of heat stored or saved in the sensible heat material (J), m represents the mass of the sensible heat material used (kg), Cp represents the specific heat of the stored material used (J/ kg K) and ΔT means the rise or fall of temperature difference measured (K). All mediums like solid liquid and gas can store sensible heat.

The TES materials are listed in **Table 1**, referred from Ref. [4], they have a high thermal capacity, are obtainable in profuse and found to be low priced. To select a suitable sensible heat storage material, the properties required have higher thermal

Material	Density (kg/m^3)	Specific heat (J/kg K)	Volumetric thermal capacity(106 J/m^{3}K)
Brick	1,800	837	1.51
Clay	1,458	879	1.28
Sandstone	2,200	712	1.57
Concrete	2,000	880	1.76
Wood	700	2,390	1.67
Glass	2,710	837	2.27
Aluminium	2,710	896	2.43
Steel	7,840	465	3.68
Iron	7,900	452	3.57
Magnetite	5,177	752	3.89
Water	988	4,182	4.17
Gravelly earth	2,050	1,840	3.77

Table 1.
Typical materials used in sensible heat TES storage.

conductivity, higher specific heat, lower density, lower vapour pressure, higher diffusivity, complementary to the storage tank materials and chemical stability. Sometimes sensible heat also needs high fluidity to carry the heat from one place to another, such as water and oil, which is also mentioned as the HTF.

1.2.2 Latent heat thermal energy storage

Latent heat TES utilises the change of phase of latent heat TES material. The transition of phase change is utilised to heat up or retrieve heat from the latent heat material by melting or solidification from solid to liquid or vice versa respectively. On melting, large amounts of latent heat are transferred at a consistently constant temperature to the latent heat material; on solidification of the material the stored heat is released. The materials utilised here are called latent heat thermal energy storage (LHTES) materials and are named as phase change materials (PCM). The quantity of heat stored can be quantified in phase change material, as shown in Eq. (2)

$$Q = m.L \tag{2}$$

Q represents the heat stored for the latent heat storage material used (J), m represents the mass of latent heat material used (kg), whereas L represents the enthalpy of the phase change (J/kg). Water is being utilised most commonly as ice for cold storage. There are other materials which are listed in **Table 2** are referred from Ref. [4]

Though numerous materials have been reviewed for PCM, only a small number of latent heat materials have been practically used, mostly due to corrosion, subcooling, separation of phase, and low conductivity of heat, longer-term cyclic stability. So generally, PCMs are chosen depending upon the suitable temperature, enthalpy required for melting, accessibility and price.

The comparison between the different hot storage media as shown in Ref. [4], the operating temperatures can be in different ranges. The specific heat and cost are less

Material	Melting temperature (°C)	Range of melting enthalpy (MJ/m³)
Water	0	330
Water-salt solutions	−100 to 0	200–300
Paraffins	−20 to100	150–250
Clathrates	−50 to 0	200–300
Salt hydrates	−20 to 93	200–600
Sugar alcohols	20 to 450	200–450
Nitrates	120 to 200	200–700
Hydroxides	150 to 400	500–700
Chlorides	350 to 750	550–800
Carbonates	400 to 800	600–1,000
Fluorides	700 to 900	>1,000

Table 2.
The melting temperatures and latent heat enthalpies of the different materials.

for the sensible heat storage materials, however, the latent heat observed in PCM's is higher and is available at constant temperature.

1.2.3 Thermochemical energy storage

The thermochemical form of energy is stored and generated once a high-energy thermochemical reaction is utilised to store up energy. Products of the resultant reactions and the heat are separately stored during the forward reaction. In the reversed reaction, the stored heat is retrieved. Thus, a reversible reaction only could be utilised for this method of heat stored.

There are two types of thermochemical energy storage, mainly bifurcated into thermochemical reactions and sorption systems. Preferred high intensity of thermal energy storage density, along with cyclic reversibility, is the prerequisite of any chemical reaction. The thermochemical energy transformation has improved the efficiency of performance than the methods adopted by physical nature. It is difficult to search for the proper chemical reaction, which is reversible, for the energy source utilised. In chemical reactions, reversibility is an important behaviour of materials required.

Higher temperatures of greater than 400°C and with higher enthalpy in the range of 80–180 kJ/mol are the properties of thermochemical materials (TCM) used for the thermochemical reactions. The reactants– products are stored separately from the heat of reactions; the TCM systems are useful as seasonal storage systems.

For example, Zeolites and Silica gels are commonly used adsorbents. To characterise these storage materials, the probable temperature lifts, the breakthrough curves, energy density and the thermal coefficient of the absorbent volume performance are to be understood as also illustrated in **Figure 2** referred from Ref. [4].

The volumetric storage capacities are illustrated in **Figure 3** in addition to a schematic comparison. Water being easily available is selected as a representation of the sensible heat storage materials taken from [1]. There are no strict limitations on the coloured region boundaries of PCMs and TCMs. It is a gross illustration of the overall

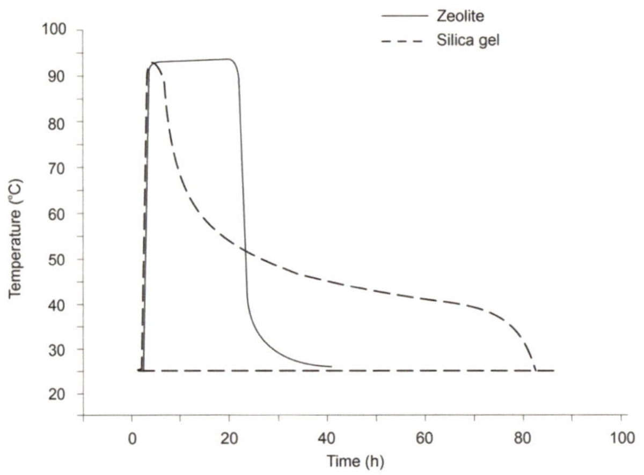

Figure 2.
Curves of thermal breakthrough (adsorption) of zeolite and silica gel.

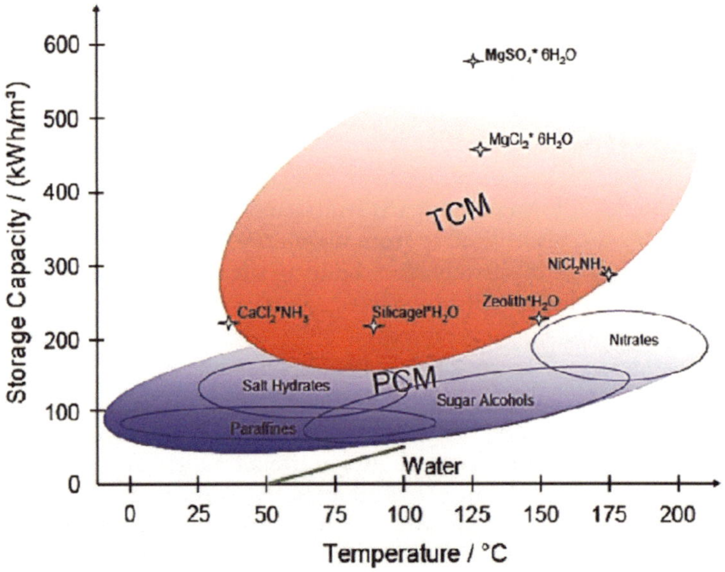

Figure 3.
Storage capacities of various PCM.

trend, and appreciably other TCMs and PCMs are available at high temperatures. A broader understanding of the available PCM with higher temperatures of melting is visualised in the following **Figure 4** taken from Ref. [5].

2. Design criteria

The thermal energy storage system consists of multiple components like the heat exchanger based on the phase change materials, the pumps, solar panels, insulations, storage tanks, etc. Each component has a different design criteria based on the aspects

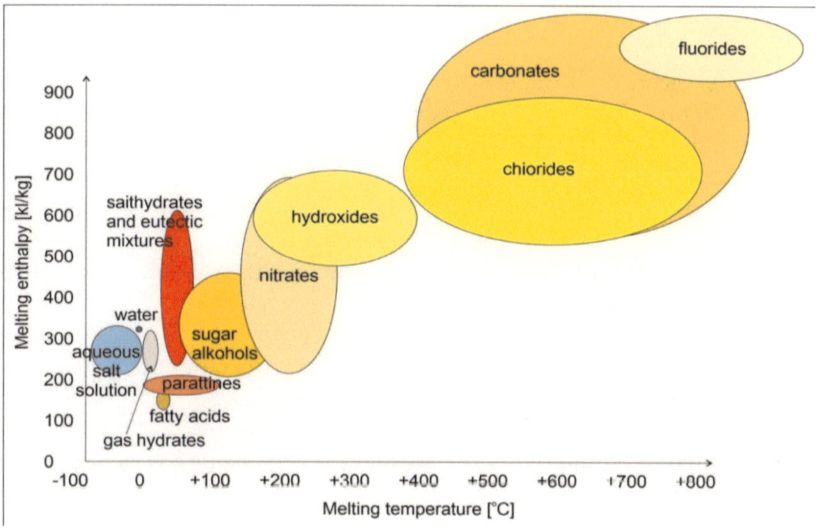

Figure 4.
Existing PCMs melting temperature vs. phase change enthalpy.

such as available heat content, flow rates, and applications. The aim of the design criteria is to maximise and optimise the heat transfer with the most effective costing using available resources and components designed for it. The established design criteria for the heat exchanger components to achieve their targets are discussed in the next section is referred from Ref. [2].

2.1 Selection criteria for thermal energy storage system

In CSP plants or any process industries, the TES system depicts an important part in the stability of generation and power supply to be met with energy demand; nevertheless there are only few TES plants with high temperature, tested using thermal energy storage and have a lot of scope for research. Thermal energy storage systems must meet the following criteria:

- High storage capacity or energy density
- Chemical and mechanical stability
- Efficient heat transfer between HTF and the storage medium
- Reversibility (maximum amount of energy must be extracted from the storage medium using HTF during charging and release this energy during discharging then return to the initial state)
- Minimum thermal losses
- Enhancement of Heat transfer can be done to the TES, if it is compatible along with the implementation of the system complexities.

2.2 PCM selection criteria

Melting point becomes the main criteria for the selection of the phase change material which in turn depends on the range of operating temperatures of the HTF. However heat transfer losses and enhanced effectiveness of the system are also necessary. The design criteria properties are differentiated into thermodynamic, chemical, economic and nucleation properties. The significant properties are listed as follows:

- High latent heat or enthalpy is produced to change the material phase at constant pressure and temperature.
- Corrosion resistance helps the material not to damage other contacted substances during high temperature and pressure working cycles.
- Environmental pollution to the society and the organisms has to be avoided by using non-toxic substances.
- High capacity of heat storage shows the material's ability to store a high quantity of energy per mass and volume.
- High thermal conductivity property ensures faster rate of heat exchange with the surroundings.
- Inflammable material ensures no burning or combustion of the material.
- The criteria also involve to avoid sub-cooling or liquefying of the PCM below its boiling point and super-cooling or solidifying of the PCM above its melting point.

The material degrading should be avoided during the process of phase change.

Considering the other important criteria required for a suitable design of PCM material are high heat density, high thermal conductivity, non-corrosiveness and economic price. In addition, the range of melting point and the availability of the PCM have to be probed.

2.3 Heat transfer fluid selection criteria

The selection of heat transfer fluid will be based on the following criteria:

- The commercial availability of the HTF and the stability of the HTF are primarily important criteria.
- The criteria of handling complexity detect the possibility of handling the HTF in the TES system.
- The criterion of phase change temperatures and vapour pressure detects whether the HTF is suitable for the TES system operating temperature and pressure.

2.4 Heat exchanger selection criteria

Selection of design criteria for the heat exchanger type is mainly prioritised as

Performance efficiency is an important criterion and it studies the affects of heat exchanger factors and parameters.

The availability and the compatibility of the components of the heat exchanger without corrosion or contamination along with the cost-effectiveness are the important criteria.

Implementation complexity criteria detect the implementation of the system complexities and handling issues.

2.5 Criteria for heat exchanger components material

In general, commercial availability, compatibility, stability and cost-effectiveness of the components investigation are few important common criteria considered for all the components and their material.

- Density is a criterion that indicates the size and portability of the design.
- High thermal conductivity criterion illustrates the heat conduction in the material preferable.

2.6 Criterion for insulating material

- As mentioned earlier, the criteria such as commercial availability, least thermal conductivity and withstand maximum operating temperature along with the cost-effectiveness for the insulating material help to appropriately choose the insulating material.

For an efficient thermal energy storage system, the selection of suitable thermal energy storage media based on different criteria and systems is essential. Basically, both sensible and latent heat storage systems are feasible. The heat transfer media for sensible thermal energy storage are selected based on the temperature at which the heat has to be supplied, for example: thermal oils, molten salts, liquid metals, concrete and sand. For latent thermal energy storage systems, water/steam and phase change materials were considered. Liquid metals as heat transfer fluid have been eliminated because of their high density and therefore their bad pump ability. Thermal oils are problematic due to their flammability. If the temperature is less than 100°C, water can be used as a heat transfer fluid.

3. Heat exchanger design

There are many types of heat exchangers that can be implemented as a TES system with PCMs but few types have been tested such as screw heat exchanger, shell and tube heat exchanger and module PCM heat exchanger. Thus the three types of heat exchangers have been discussed in the next section.

3.1 Screw heat exchanger

Screw heat exchangers (SHE) are mostly used for industrial purposes as shown in **Figure 5**. In general, it consists of a drum and rotating hollow shafts (screws) arranged

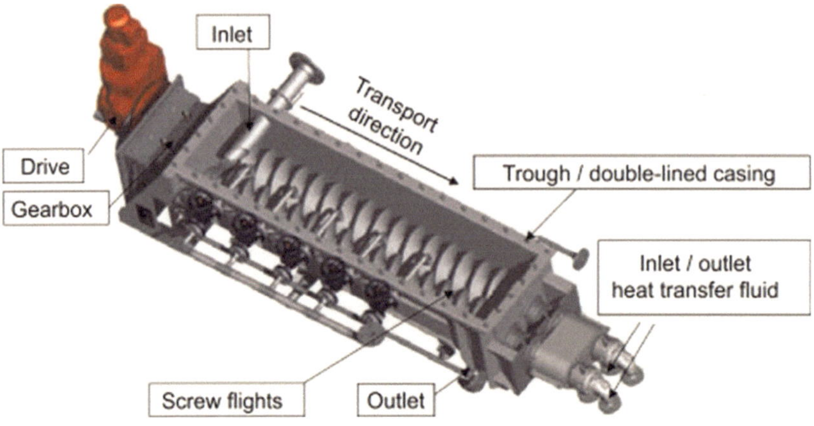

Figure 5.
Screw heat exchanger schematic.

parallel to each other, and fluid passes through them. A prototype of SHE with PCM was implemented and tested to be integrated into the CSP plant as a TES system. The PCM is placed inside SHE; the phase change takes place when HTF flows through the hollow shaft during the charging/discharging process. In this prototype, the storage system type is two storage tanks, where one is for molten PCM and the other is for solid PCM. The continuous rotation of the hollow shafts will decrease the solidification of the PCM during the discharge process. However, during the discharging process, the PCM will be solidified on the shaft, then it will be crushed due to the shaft rotation. Furthermore, its size will be large to be reused for the charging process. Therefore, a granulate crusher is required in order to decrease the size of the collected PCM, and SHE is compatible with the usage of PCMs and HTF. While the implementation is kind of complex due to the two-tank storage system, PCM granulates and the requirement of a granulate crusher. From a cost-effectiveness point of view, SHE cost depends on the size of a small prototype. The benefit of it is the alternative usage in the charging/discharging process, self-shaft cleaning due to the continuous rotation and the less solidification of PCM effects on heat transfer. However, the cost of transportation of SHE, supplied power, two-tank storage system and the crushed granular is much higher than the obtained benefits. Finally, there are no data mentioned about the efficiency of using SHE with PCMs but in general heat exchangers have a storage density of 95% of volume.

3.2 Shell and tube heat exchanger

Shell and tube heat exchanger (STHE) generally consists of one or more round tubes mounted in a parallel configuration to a cylindrical shell shown in **Figure 6**. It is the most investigated type of heat exchanger with PCM. Also, it has a simple structure and can be easily manufactured beside the variety of configurations that can be made. Additionally, PCM can be easily replaced at the end of its lifetime. STHE exchanges heat through the pipe walls between flowing HTF in pipes and the PCM contained in the shell. As HTF passes through the pipes, phase change in PCMs will occur during the charging/discharging process. In discharging, the solidification of PCM will start at the pipe walls so it will act as a thermal resistor.

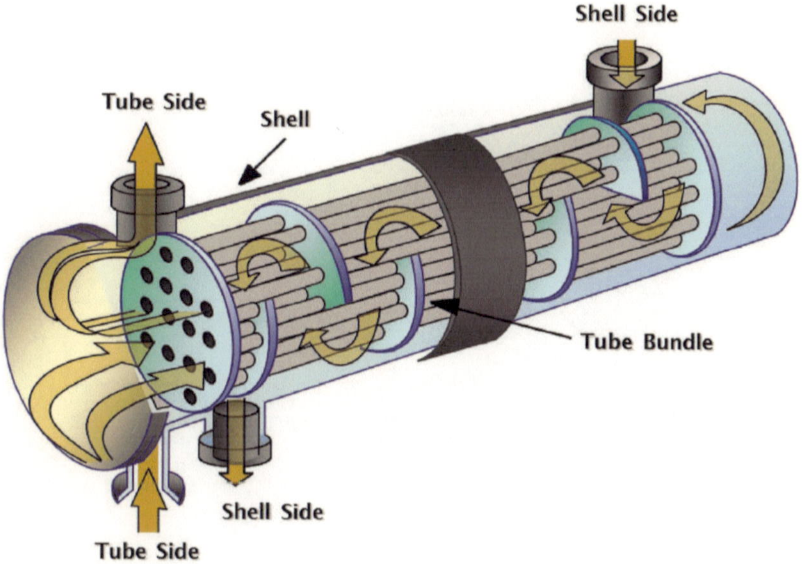

Figure 6.
Shell and tube heat exchanger.

However, the thermal energy storage system with shell and tube heat exchangers is one of the most promising and cost-effective heat exchangers for latent heat storage. Moreover, its performance was investigated in different heat transfer enhancement techniques such as fins and cascaded PCM. Therefore, available data can be used. Additionally, this design has been recognised to be the promising configuration of a latent storage system acquiring high efficiency and minimum volume. STHE can be manufactured based on design requirements. Moreover, it is compatible with the usage of HTF and PCMs as well as being integrated as a TES system. Furthermore, it has a simple structure that can be easily implemented and handled. Its cost is a function of the outer surface area. In terms of cost-effectiveness, STHE can be alternatively used for charging/discharging, heat transfer enhancement techniques can be integrated with it, and it stands as a TES system. Therefore, an extra storage tank is not needed as the space required is designed properly, and the benefit of thermal stratification can be taken. The only additional cost required is for a pumping system for fluid circulation. Finally, from performance efficiency perspective heat exchangers when used with PCM has a storage density of 95% of volume.

3.3 Heat exchanger module (HEM)

This type is similar to any heat exchanger in construction. The only difference is that Macro encapsulated PCMs are placed in modules (cylindrical or spherical or any container) where the HTF flows over the PCM modules through the tank as shown in **Figure 7**. This type of design solves issues related to volume change, heat transfer and material compatibility besides availing flexibility and high package density since it can be fabricated in different sizes and shapes. Also, modules are easily handled and shipped. The encapsulation salt particles are more effective than heat exchangers with lower possibilities of success. However, this technology is researched a lot with great potential of energy storage for the purpose of high and medium-temperature storage

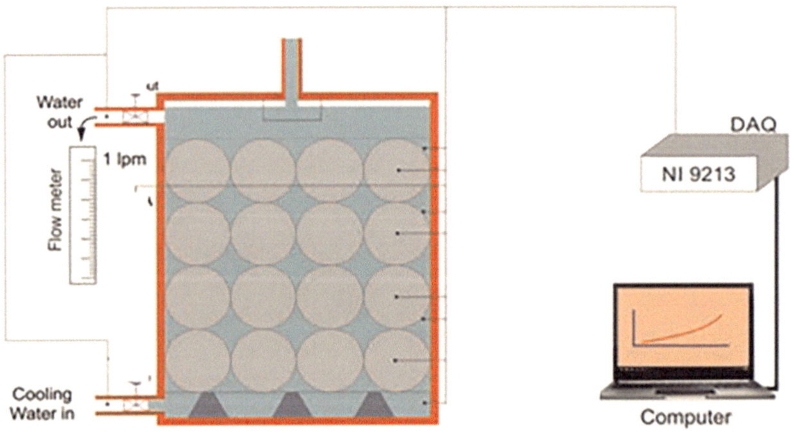

Figure 7.
Heat exchanger module (HEM).

systems. Additionally, the coating material of the encapsulation may cause contamination of HTF. In most of the cases, the replacement of PCM at the end of the life cycle is not possible, HEM is not commercial yet, therefore, they have to be individually designed. Moreover, it is not fully compatible with HTF, since the coating material may contaminate it and cause degradation. Implementation complexity is not a big issue since it is a similar and common type of heat exchanger.

In terms of cost-effectiveness, there are no common available data that estimate the cost of it but it is more cost-effective compared to heat exchangers. From a performance efficiency perspective, it has a storage density of 74%. However, heat transfer enhancement methods are being researched.

3.3.1 Design considerations for stratified HEM tanks

The following design criteria help in the design of an efficient TES tank with enhanced thermal stratification [1].

Geometrical considerations: A deep water-storage container is desirable to improve thermal stratification. The water inlet and outlet should be installed to produce a consistent water flow to evade mixing. The location of inlet and outlet openings should be placed as close as possible to the top and bottom respectively of the stored volume to minimise the dead water volume and the surface area in contact with the water.

Operating considerations: The difference of temperature linking the top and bottom parts of the tank should at the least be 5°C to 10°C. Controls can be used to maintain fixed water temperatures in the tank's upper and lower parts if desired. The velocity of the water flowing into and out of the tank should be low.

Other considerations: The insulating and water-proofing characteristics of the tank should be designed to meet appropriate specifications.

For proper installation and control, using state-of-the-art equipment requires less energy for heating and cooling with substantial potential to substitute with low initial costs. More sophisticated thermal-design calculations are required to get the best design for better operating conditions.

3.3.2 Degradation of thermal stratification

Degradation of thermal stratification or mixing in a thermal storage tank can be essentially attributed to the following physical mechanisms:

- Diffusion of heat to the bottom of the storage tank due to the temperature difference between the fluids in the top and bottom regions.

- Heat loss to the surroundings due to improper insulation.

- Effect of flow turbulence due to incoming /exiting flow rate.

- Thermal conduction along with the height of the storage tank through the walls of the tank.

The medium used for heat transfer interchanges heat through direct and indirect contact with the medium used for storage which forms a thermocline in HTF. The dual medium concept has a temperature drop as a drawback when compared to the single medium hot or cold tank while charging and discharging. The single medium maintains a constant outlet temperature while charging and discharging till the heat is removed in the tank, however, the HTF outlet temperature increases as it is discharged and decreases the more it is discharged for a dual medium storage system which leads to unfeasible storage capacity.

During the charging and discharging periods, mixing forms the main cause of stratification loss in general and major mixing losses occur during the lengthy storage periods in general. Enhancing the stratification significantly increases the efficiency of TES compared to a thermally mixed-storage tank. Hence the TES tank has to be evaluated quantitatively and comprehensively to clearly analyse the effects of stratification on the performance of a TES system.

3.3.3 Parameters to measure the stratification

Different methods for the stratification characterisation in TES system are being recommended over the years. A density method and a temperature approach are two fundamentally distinct approaches to stratification. In general, only temperature-based techniques are presented. Furthermore, it is critical to distinguish between elements that effect stratification (e.g., Richardson number, H/D ratio, etc.) and figures that evaluate the degree at different time intervals. Furthermore, the thermocline thickness, which divides the hot and cold zones inside the storage, may be used to analyse indicators of the degree of stratification.

Numerical figures given in terms of efficiency or ratio, on the other hand, are typically used for comparing the experimental storage process to a hypothetical storage process and thus include information about the history of the storage process. The most commonly used dimensionless numbers, such as MIX number, Richardson number, ratio H/D, discharging efficiency ratio, Peclet number and Reynolds number, to characterise stratification in water tanks and studied their suitability.

According to their findings, Peclet and Reynolds values do not explain stratification inside the tank, and it is unclear if discharging efficiency can define stratification inside the storage tank at low flow rates.

Richardson number is used to evaluate the stratification by comparison, for example, between PCM-filled storage against a reference case without PCM. The Richardson number is a popular way to describe stratification. It is a ratio of buoyancy forces to mixing forces; a low Richardson number implies a mixed storage tank, whereas a high Richardson number indicates a stratified storage tank.

The MIX number is used to analyse stratification in a water storage tank at a specific time, with a figure ranging from 0 to 1 indicating the degree of stratification; nevertheless, it is sensitive to slight variations in the temperature profile. Because the stratification degree is of considerable importance in this work, several stratification efficiencies such as energy efficiency (first law of thermodynamics), exergy efficiency (second law of thermodynamics), stratification efficiency and MIX number were utilised. The MIX number ranges from 0 to 1, with MIX = 0 representing a perfectly stratified tank and MIX = 1 for a fully mixed tank.

Stratification efficiencies based on the first law of thermodynamics most commonly compute a proportion of energy recovered with a certain charging and discharging operation with a set intake temperature and mass flow. The charging energy efficiency is defined by Ref. [4] and is based on the first rule of thermodynamics.

Methods for evaluating exergy efficiency based on the second law of thermodynamics are especially relevant when the energy will be utilised to do work, but they are not the only ones. The exergy stored in this example represents the thermodynamic limit of the work that may be generated. To calculate values based on exergy, a reference' dead state' that corresponds to the thermodynamic condition of the environment must be determined. Exergy is always greater in storage with a more significant temperature gradient and better stratification than in storage with an identical energy content but less pronounced stratification. As a result, numbers based on the second law of thermodynamics may be utilised to provide information regarding stratification efficiency.

Oro et al. [6] and Cabeza et al. [7] reported that the various parameters such as degree of stratification first law efficiencies, second law efficiencies, mix number, stratification number and Richardson number were used to characterise the storage system and they reported that energy and exergy efficiencies have no relationship with thermal stratification.

3.3.4 Advantages of stratification

Several studies have illustrated that the stratified STES system will give better results in comparison to a mixed tank of STES. In a STES system when connected to a solar panel, the heat extracted from a stratified charged tank is greater and for a longer duration in contrast to a fully mixed STES system. This stratification causes the low temperature at the bottom of the fluid of the tank returning to the solar collector and this creates a higher temperature difference between the fluid and the collector. It is concluded that the use of stratification in thermal storage designs should be considered as it increases the exergy storage capacity of thermal storage.

Developments in Stratification Analysis Stratified TES tanks often exhibit superior performance to non-stratified tanks, especially in low HTF flow rate heating systems. The level of TES stratification is decreased by HTF mixing, due to natural or forced convection, or by heat diffusion by conduction. Haller et al. [3], reviewed and compared methods to determine the stratification efficiency of a TES during a complete thermal cycle. Only one method considers the entropy generation, which provides a more meaningful interpretation of efficiency than energy analysis. Some of the advantages of stratified thermal storage systems in construction are as follows:

- They can often incorporate inexpensive above-ground steel tank construction.
- They provide a less expensive option than adding more chillers for meeting peak cooling loads.
- They are normally compatible with any water chiller technology.
- They can be incorporated into air conditioning systems, process cooling or heating and district cooling or heating loads.

Research activities on TES are ongoing at various national laboratories, universities, and research centres worldwide. The TES system can utilise thermal stratification. Such stratification separates lower-density warm water above higher-density cool water and can be applied in a single economic storage tank to provide demand-side management of cooling or heating loads. Stratified thermal storage is expected to minimise the tank's costs, insulation, system integration, piping, and controls. Consider the example of cool supply water being withdrawn from the storage tank's bottom during peak cooling periods and used directly in the cooling loop and then sent back to the top of the tank. However in the off-peak periods, warm water is taken from the tank's top and cooled with the help of low-cost off-peak energy, and sent back to the tank's bottom.

4. Experimentation on the heat exchange module

An experimental project was conducted using a HEM and its experimental setup [8, 9], along with all the design data is being discussed in the following sections.

4.1 Experimental set-up

The experiment represented in **Figure 8** has thermal energy storage with a heat exchanger module or storage tank, a circulating hot water bath, a electric heater to simulate a solar heat situation, components of a heat transfer loop along with a data acquisition system. A cylindrical storage tank was fabricated with a height of 300 mm and a diameter of 320 mm as shown in **Figure 8a**. A 50 mm space on the top and a 25 mm space on the bottom is the working space left free and in between these two regions a packed bed of PCM balls is arranged in four layers and each layer was distinguished using a steel mesh and had 11 PCM balls in each layer which were dispersed uniformly within the layers of the storage HEM module. The PCM HS89 was a phase change storage medium and is a hydrated salt having a melting temperature of 89 °C which was procured from Pluss advanced technologies Pvt. Ltd., New Delhi. The properties such as specific heat, phase transition temperature, liquid and solid densities and the latent heat were got from the technical data sheets of the company and are presented in **Table 3**. The 2 mm thick stainless steel encapsulated spherical ball has an average outer diameter of 80 mm. A 40 mm thick insulation of polyurethane foam for the storage tank was used to minimise the heat loss.

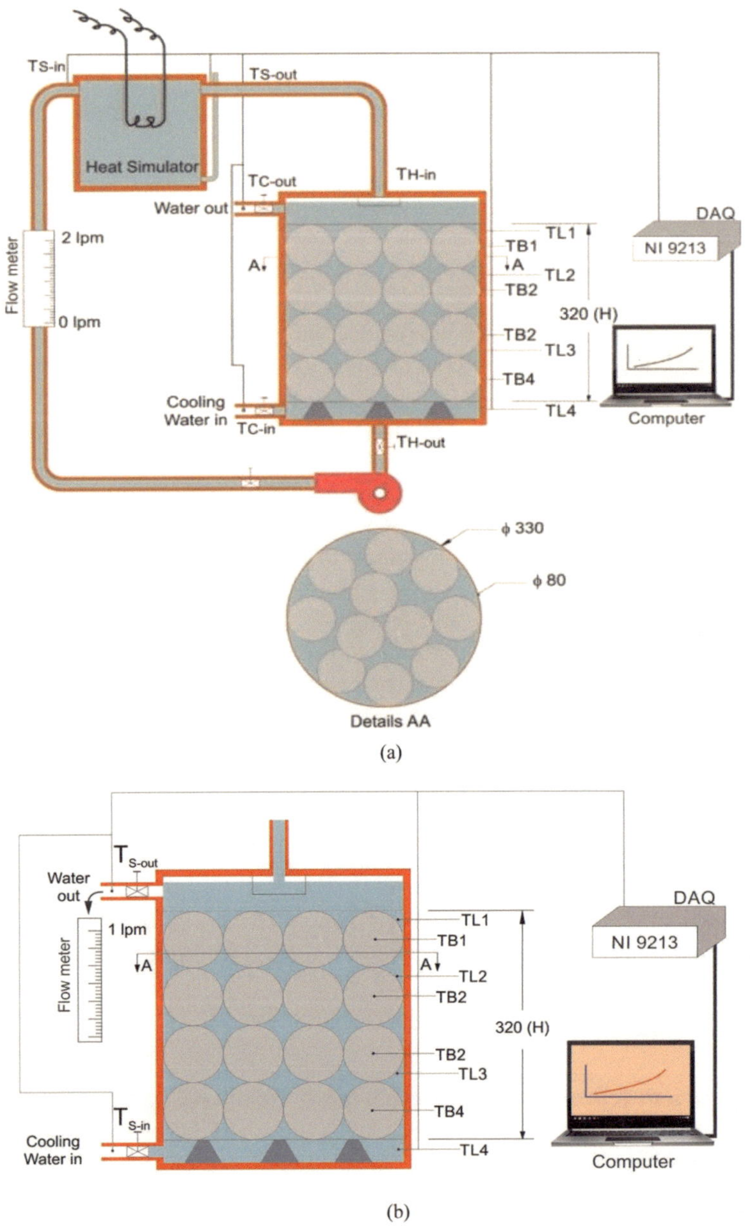

Figure 8.
Charging and discharging of a HEM and its circuit illustration (a) charging experiment with storage tank (b) discharging experiment.

The 30 liter HTF hot water bath is simulated with two immersion heaters of 1.5 kW capacities. The temperature of the inlet water to the storage tank from the heater gradually increases until it reaches 95°C, and this temperature was controlled with the help of a thermostat. To observe the water level as lot of steam evaporates the water in the heater tank, a glass tube was parallelly attached to the HEM as shown in **Figure 8**. The steam produced while heating the HTF or water reduces the level of the water and as the level reduces below a required liquid level, makeup water is added. A quarter-watt centrifugal

S. No	Property	Value
1	Melting temp (°C)	87–89
2	Liquid bath latent heat (kJ/kg)	125
3	Liquid bath freezing temp (°C)	89
4	Solid density (kg/m^3)	1630
5	Solid specific heat (kJ/kgK)	2.65
6	Liquid density (kg/m^3)	1540
7	Melting temp (°C)	2.65

Table 3.
Technical specification for HS (Hydrated Salts) 89 PCM.

pump is used to circulate the HTF from the storage tank to the hot water bath and if the water level was increased beyond the brim of the hot water bath it would flow to the HEM through gravity. A rotameter is placed in the reverse flow from the storage tank to the hot water bath to measure the flow rate and maintain it to 1 liter per minute using a valve placed at the bottom of the HEM tank. The inlet and outlet pipes are placed separately in HEM and used to circulate the water based on the charging or discharging applications as illustrated in **Figure 8**. The temperature in the PCM encapsulation in each layer was measured using a thermocouples located inside any one of the PCM balls itself. The thermocouples were inserted in the head of the PCM balls by drilling a hole in the nut of the encapsulation, inserting the thermocouple, and then applying an adhesive (m-seal-curing epoxy compound) to avoid leakage. Four thermocouples are placed equally among the layers of water to measure and record the layer temperatures. Four thermocouples are also placed in the encapsulated PCM balls to record the PCM temperature inside the ball to identify whether the PCM has melted on heating with HTF. A data acquisition system NI 9213 was used to connect the thermocouples to the laptop using LABVIEW interface. The data is recorded in MS-EXCELL sheets for further analysis.

4.2 Experimental procedure

In the present experiment, initially after checking all the connections of the circuit charging of the HEM was conducted by allowing the heated water to flow from the hot water bath to the HEM across all the balls so that the PCM capsules get heated and reach the temperature of 89°C. As the temperature of the water increased beyond 90°C, the PCM in the spherical capsules is melted, and while melting it is simultaneously ensured that the hot water bath did not go beyond 95°C using thermostat. Using a centrifugal pump the water from the bottom of the HEM is pumped back at a rate of 1 L/min to flow to the hot water bath, so the centrifugal pump was switched on. After reaching the melting temperature, the temperature of the balls remains constant due to phase change taking place in the HS89 salts in the PCM balls. The data acquisition system NI 9213 is used to monitor and record continuously the increase in temperatures with respect to time. As the temperature of the balls reaches 89°C and remains constant, salts in the PCM balls will melt. This constant temperature indicates the charging of the balls and the HEM is completely charged. After recording the temperatures of balls and the layers of the charging experiment using thermocouples and DAQ as shown in **Figure 8a**, the data is stored and the charging experiment is complete. After ensuring the finish of charging of the experiment as the complete phase change takes place, the discharging experiment was initiated. In the discharging experiment, the centrifugal pump was stopped and the hot water supply was cut off. Room temperature water at 30°C is circulated from the cold water inlet at a rate of 1 L/min into the hot-charged HEM. As the thermocouples placed inside the balls and within the layers start recording the temperatures at various positions while discharging. The outlet temperature of the water coming out from the storage tank also was recorded separately during this process of discharging. Instantaneous and cumulative heat transfer, charging and discharging efficiencies and the analysis of stratification characteristics for the storage tank were deduced from the transient temperature variation obtained in the PCM and the HTF of the HEM.

4.3 Data analysis

The various equations used to evaluate the performance of the storage tank such as Q_{ins}, Q_{cumm}, Q_{loss}, Q_{stored}, charging efficiency, stratification number and Richardson number are presented in this section. The analysis of the experiments can be done by deriving these parameters which help us to identify the different aspects of the heat flow and the stratification behaviour across the storage tanks. The analysis is drawn based on the values of these parameters also helps to conclude which storage tank is better for which purpose.

4.3.1 Overall heat loss coefficient ($U_{overall}$)

This parameter is evaluated from the drop in temperature of the water in the storage tank over a long duration of time when the storage tank is under idle condition (without the PCM balls) using Eqs. (3) to (6). This parameter helps to calculate the heat loss and becomes important to understand the loss of heat while charging and discharging experiments are conducted:

$$Q = U_{overall}.A.T_{LMTD} \tag{3}$$

$$U_{overall} = \frac{Q}{A.T_{LMTD}} \tag{4}$$

$$Q = mC_p\left(T_{ini} - T_{final}\right) \tag{5}$$

$$T_{LMTD} = \frac{\left(T_{ini} - T_{\infty}\right) - \left(T_{final} - T_{\infty}\right)}{lnln\frac{\left(T_{in} - T_{\infty}\right)}{\left(T_{final} - T_{\infty}\right)}} \tag{6}$$

Q = Total heat lost to the ambient, W
T_{ini} = Initial temperature in the tank, °C
T_{final} = Final temperature in the tank, °C
T_{∞} = Ambient Temperature, °C
Cp = Specific heat of water, kJ/kgK
m = mass of water in the idle tank, kg
A = outside area of the idle tank, m^2

The overall heat loss coefficient ($U_{overall}$) is evaluated for all three storage tanks and the average value is considered for further analysis.

Separate experiment was conducted under idle conditions to evaluate the average heat loss coefficient. The temperature of the water in the storage tank was averaged out for all the layers of water with respect to time. The initial and final temperatures of the layers while discharging experiments under idle conditions were taken to evaluate the LMTD as given in Eq. 6. The overall heat loss coefficient was calculated with Eq. 4 and an average value of 5.056 W/m^{2}K was being evaluated for the storage tank and used for calculations of both charging and discharging of the storage tank with PCM balls

4.3.2 Instantaneous heat transfer (Q_i)

The instantaneous heat transfer represents the amount of heat transferred into the storage tank at any instant of time during the charging process as given in Eq. (7). It shows the effect of heat based on the inlet and outlet temperature and is directly proportional to the stratification of the tank.

$$Q_i = \dot{m}C_p(T_{H-in} - T_{H-out}),\text{W} \tag{7}$$

T_{H-in} = Inlet HTF temperature to the tank, °C
T_{H-out} = Outgoing HTF temperature from the tank, °C
$\dot{m}$ = Mass flow rate of the HTF, kg/s

4.3.3 Cumulative heat transfer (Q_c)

The cumulative heat transfer represents the total amount of heat stored at any instant from the start of the charging process as Eq. (8) given below shows. It adds on to the total heat from the starting of the experiment and this cumulative heat indicates the total amount of heat either stored while charging or retrieved while discharging of the storage tank

$$Q_c = \sum_0^t \dot{m}C_p(T_{H-in} - T_{H-out})_t \Delta t,\text{W} \tag{8}$$

t = Instantaneous time

4.3.4 Heat lost (Q_{loss})

It is the amount of heat lost from the external surface of the storage tank during the charging process. It is evaluated using Eq. (9).

$$Q_l = U_{overall}\, A\, dT \qquad \text{W} \tag{9}$$

$dT = T_{avg} - T_\infty$
T_{avg} = Average temperature of all the layers of water in the storage tank, °C
$U_{overal\ l}$ = Overall heat loss coefficient of the storage tank, W/m^{2}K
A = Area of the storage tank, m^2

4.3.5 Heat stored (Q_S)

It is the actual amount of heat retained in the storage tank during the charging or discharging process after subtracting the heat lost to the surrounding, from the storage tank. It gives an idea as to what amount of heat is stored and is calculated using Eq. (10).

$$Q_{stored} = Q_{cum} - Q_{loss} \tag{10}$$

4.3.6 Charging efficiency

The charging efficiency of the storage tank is the ratio of the instantaneous heat transfer to the maximum heat transfer at a given inlet temperature at any instant, keeping constant the flowrate of the HTF at the inlet. It illustrates how the efficiency varies with respect to time for the charging done to the storage tanks without and with PCM. It is determined by the following Eq. (11)

$$`\eta_{ch} = \frac{[T_{H-in} - TLb]_t}{T_{H-in} - T_{ini}} \tag{11}$$

T_{ini} = Initial temperature of the tank

4.3.7 Discharging efficiency

The discharging efficiency of the storage tank is the ratio of the instantaneous heat transfer to the maximum heat transfer at a given inlet temperature at any instant keeping constant flowrate of the HTF at the inlet as given by Eq. (12). The cold water inlet at 30°C (TC-in) is made to enter the storage tank till the end of the experiment.

$$\eta_{disch}(\mathrm{t}) = \frac{(T_{C-out} - T_{C-in})_t}{(T_{C-out})_{ini} - T_{C-in}} \tag{12}$$

T_{Cin} = Cold inlet water temperature to the tank at the bottom, °C
T_{Cout} = Outgoing water temperature at the top, °C

4.3.8 Stratification number (Str)

It is the ratio of the average temperature gradients at each interval of time to the temperature gradient at the initial charging process (t = 0). Eq. (13) shows the evaluation of stratification number, depending on the temperatures acquired at equi-distant points in the thermal energy storage tanks

$$Str = \frac{(\partial T/\partial y)t}{(\partial T/\partial y)t = 0} \tag{13}$$

where

$$\left[\frac{\partial T}{\partial y}\right] = \frac{1}{N-1}\left[\sum_{k=1}^{N-1}\frac{(T_k - T_{k+1})}{\Delta Z}\right]$$

$$\left[\frac{\partial T}{\partial y}\right]_{t=0} = \frac{T_{H-in} - T_{ini}}{(N-1)\Delta Z}$$

where k is the nodal points where the temperature measurements are made, N is the number of nodal points and ΔZ is the distance between the nodal points $T_{\_(H\text{-}in)}$ and $T_{\_ini}$ which shows the hot water inlet and initial temperatures of the HTF, respectively.

4.3.9 Richardson number (Ri)

It is an effective indicator of stratification performance that considers the various aspects of storage systems such as the geometry of the storage tank, specific velocity, and the top and bottom temperature of the layers within the tank. It is given by the ratio of the buoyancy forces to the mixing forces and is estimated by Eq. (14).

$$\mathrm{Ri} = \frac{g\beta H[(T_{H-in}) - (T_{H-out})]_t}{v_{sf}^2} \tag{14}$$

$$\nu_{sf} - \frac{V}{\pi r^2}$$

where $T_{H\text{-}in}$ and $T_{H\text{-}out}$ represent the top and bottom temperature of the HTF measured in the TES tank and H is the distance between these locations. V and ν_{sf} represent the volumetric flow rate and superficial velocity of HTF entering the TES tank, respectively and r is the radius of the inlet pipe which is ¼th of an inch in the present experiment.

4.3.10 Mix number

Mix number is the level of energy in the TES tank weighted by its vertical distance from the bottom of the storage tank. The vertical moment of energy in a storage tank was shown in Eq. (15).

$$M_E = \int_0^H y dE \tag{15}$$

This equation was further deciphered as shown in Eq. (16), as follows

$$M_E = \sum_{i=1}^{N} y_i E_i \tag{16}$$

where N = the number of layers used

y_i = the distance measured from the bottom of the tank.

$E_i = (\rho C_p V_i T_i)_{water} + (\rho C_p V_i T_i)_{pcm}$, which is the energy stored at a particular level i

On calculating the energy stored for PCM, the sensible heat is accounted as $(mC_p V_i T_i)_{pcm}$ and the latent heat is accounted as $(\rho V_i L)_{pcm}$ cumulatively based on the temperature which the PCM attained on heating or cooling and when the PCM reached the phase change temperature. The term V_i represents the volume of the PCM or the water at a particular level i and L here represents the latent heat of the PCM when the PCM reaches the phase transition temperature.

When considering mixing as a function of time of the day, the moment of energy increases with increase in temperature of the heated water and decreases when hot water is withdrawn from the tank. The temperature profiles were determined experimentally in the fully mixed and the unmixed storage tank. Hence the moments of energy for the unmixed or fully stratified (M_{str}) and the fully mixed (M_{mix}) are determined and substituted in the *MIX* number as shown in Eq. (17) given as follows:

$$MIX = \frac{(M_{str} - M_{actual})}{(M_{str} - M_{mix})} \tag{17}$$

For a given set of inlet conditions, the M_{str} and M_{mix} represent the largest and the smallest values of the moment of energy, respectively.

4.4 Error analysis

The error associated with various primary experimental measurements and the calculation of estimated uncertainties for the performance parameters are given in **Table 4**.

4.5 Results

The temperatures being measured across the HEM tank along with the PCM while carrying out the experiment are plotted under various dimension with respect to time and analysed in the following section. The temperature variation is initially noted with respect to time and later on various types of data like the amount of heat transferred is quantified along with the different stratification analysis.

Fundamental parameters	**Uncertainty (%)**
Mass (m)	1
Length (L)	0.253
Diameter (D)	0.4167
Volume (v)	1
Temperature (T)	0.334
Time (t)	0.033
Derived parameters	
Mass flow rate ($\dot{m}$)	1.054
Volume flow rate ($\dot{v}$)	1.054
Instantaneous heat rate (Qinst)	1.053
Cummulative heat rate (Q cumm)	1.0548
Overall heat loss coefficient (Uoverall)	1.3853
Heat lost (Qlost)	1.6173
Heat stored (Q stored)	1.9309
Charging efficiency (ηch)	0.4723
Stratification number (Str)	0.5927
Richardson Number (Ri)	1.7268

Table 4.
Summary of the estimated uncertainties.

4.5.1 Comparing the charging and discharging characteristics of the storage tanks filled with PCM for all the HEM

Figure 9 shows the temperature variations of the heat transfer fluid in the top and bottom layers of the storage tank along with the variation of the HTF inlet temperature during the charging process. **Figures 9a** and **b** represent the temperature variations of the HTF in the storage tank with PCM balls while charging and discharging respectively. In both the figures, the shaded portion represents the temperature variations of the later of HTF along the height of the storage tank at any instant of time. The thick line on the top for **Figures 9a** and **b** represents the incoming and outgoing HTF temperature of the storage tank on charging and discharging, respectively.

In the case of storage tank with PCM balls, while charging as shown in **Figure 9a**, the incoming hot water referred by the dark line on the top of the storage tank is slowly heated and the temperature of it rises until it reaches around 95 °C after which the thermostat stops heating the incoming water temperature to rise, as it forms steam at 100 °C. As the cold water is circulated from the bottom back to the heater tank, the HTF temperature keeps rising due to the mixing of incoming hot water and the layers temperature slowly rises. Hence the stratification or the temperature difference for the HEM continuously increases to 16.68°C till the end of charging upto 270 minutes. This temperature difference of more than 15°C is maintained for more than 125 minutes at the end of charging.

However in discharging we find that a maximum temperature difference above 20°C is available for 25 minutes and it diminishes to 7°C at the end of discharging upto 66 min. The $T_{c\text{-}out}$ temperature of water is the outgoing hot water temperature and more than 80°C of water can be retrieved for around 24 minutes and later on the temperature of the outgoing water reduces gradually. The outgoing water temperature runs through the center of the shaded portion as mixing of the HTF layers are predominant. Substantial number of balls present with high heat capacity is facing the

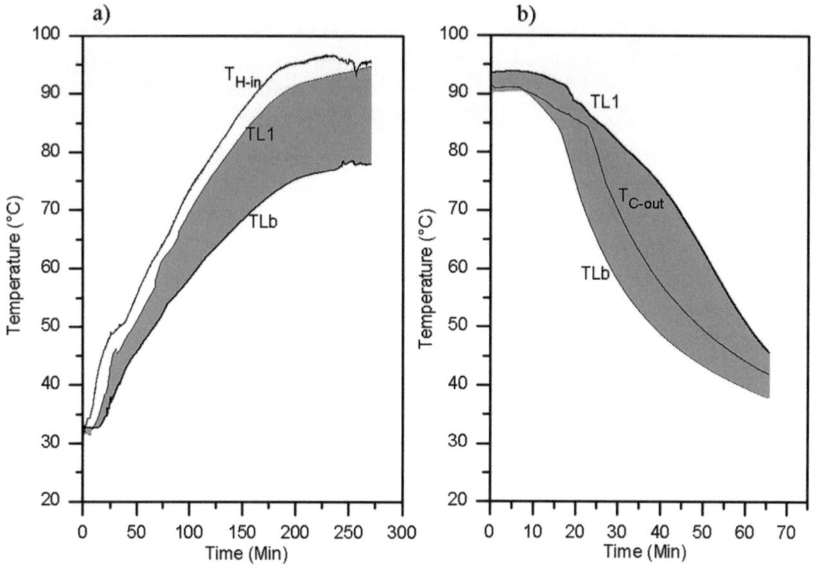

Figure 9.
Temperature variation in the top and bottom layers of the storage tank with PCM balls in the HEM while a) charging b) discharging.

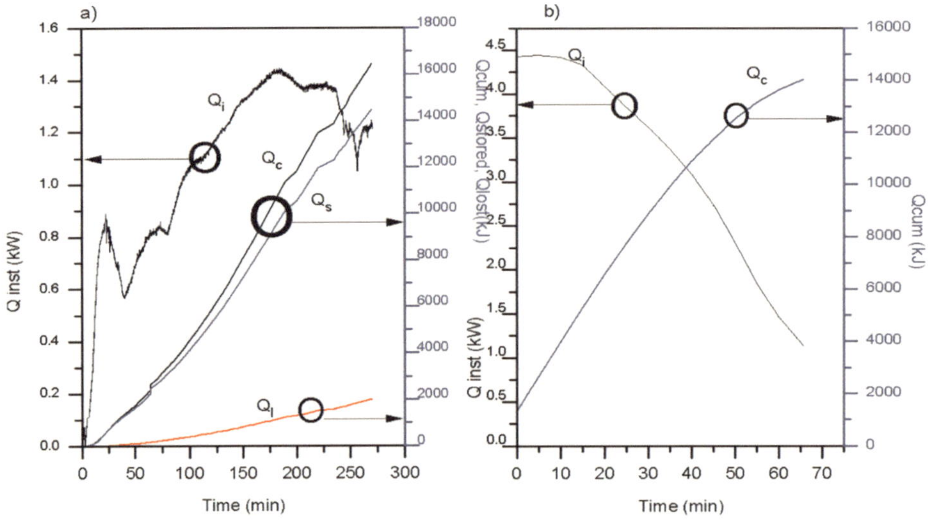

Figure 10.
Instantaneous and cumulative heat transfer during the charging and discharging process.

incoming hot water in the layers of HEM, and due to this arrangement of the PCM in the storage tank, it helps to achieve the required type of stratification.

Figure 10 shows the instantaneous and cumulative heat transfer during the charging and discharging process which are calculated using the temperature data. It is seen from the figure that the instantaneous heat transfer continuously rises and the cumulative heat stored in the HEM at the end of the charging process is around 14,000 kJ. After deducting the heat lost during the charging process from the cumulative heat transfer, the heat stored is also calculated. From the graph, the heat lost is found to be significant at the end of the charging process.

During the discharging process, the instantaneous heat transfer is initially very high in the range of 4 to 4.5 kW. This higher instantaneous heat transfer is maintained for the first 10 min duration. Even at the end of the discharging, the instantaneous heat transfer is maintained consistently higher till the heat is discharged in the HEM. The gradual fall in the instantaneous heat transfer could be due to the slow retrieval of heat in the PCM at the bottom of the storage tank, due to high stratification created as the cold water enters from the bottom and cannot move up quickly due to high density. Hence in this case of the HEM, almost all the heat available in the storage tank of 14000 kJ is removed within the duration of 65 min. Since the instantaneous heat transfer rate decreases, further removal of heat from the storage tank appreciably will be a prolonged process.

Charging efficiency and discharging efficiency have been plotted in **Figure 11** for the HEM considered in the present investigation. It is seen from **Figure 11a** that charging efficiency is high initially as the temperature difference between the incoming hot water and the room temperature is high and hence stratification effect increases. Also during the charging process, the presence of reasonable quantity of PCM balls in the top layer retains more amount of heat at the top layer of HEM and hence the charging efficiency remains steady up to 38 % till the end and does not drop further.

During the discharging also the discharging efficiencies gradually drops to 20% for a time span of 66 min as the convection heat gradually flow from the top and bottom

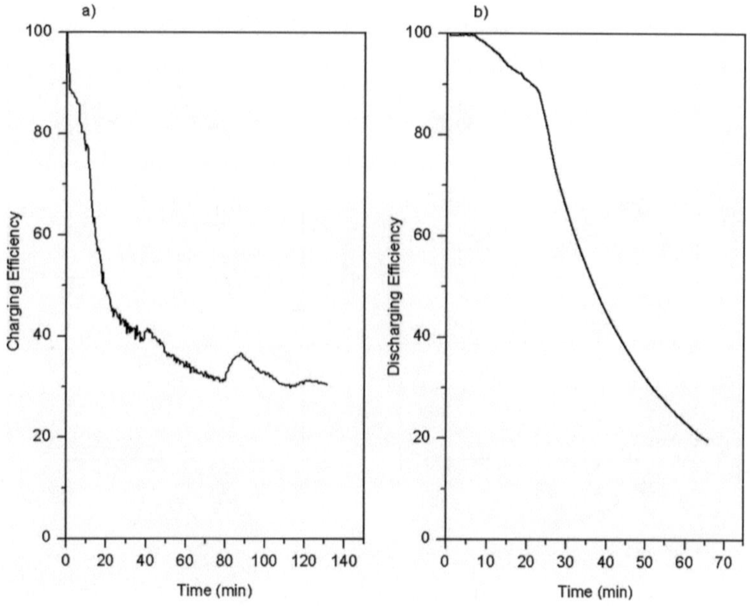

Figure 11.
a) Charging efficiency and b) Discharging efficiency for the HEM.

layers. Initially, the heat transfer is arrested as the PCM balls melt upto the first 23 min and the temperatures also remain high. This temperature pattern of discharging provides favourable heat transfer during the entire discharge process.

Figure 12 shows the variation of the stratification number in the HEM during the charging and discharging process. It is observed that in this module there is a gradual increase in the stratification that favoured the rate of charging and hence the charging efficiency is upto 40%. The stratification increases upto 160 min as there is continuous difference of temperature which causes the convection of heat from top to bottom

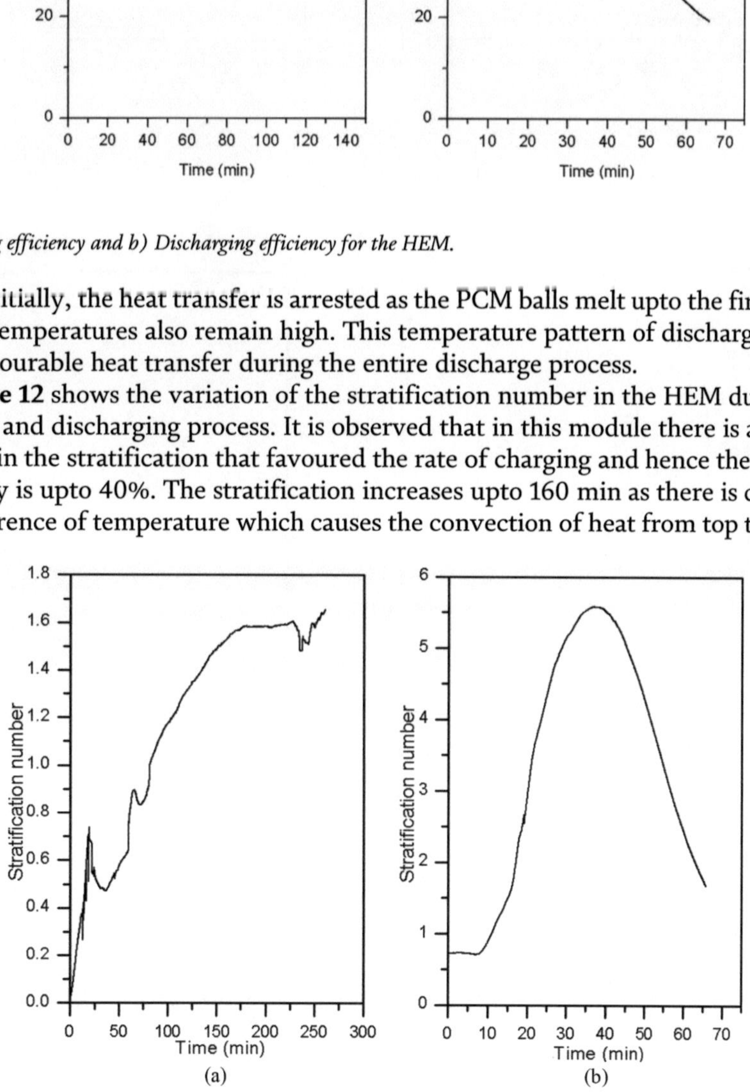

Figure 12.
The variation of stratification number for a) charging and b) discharging of HEM.

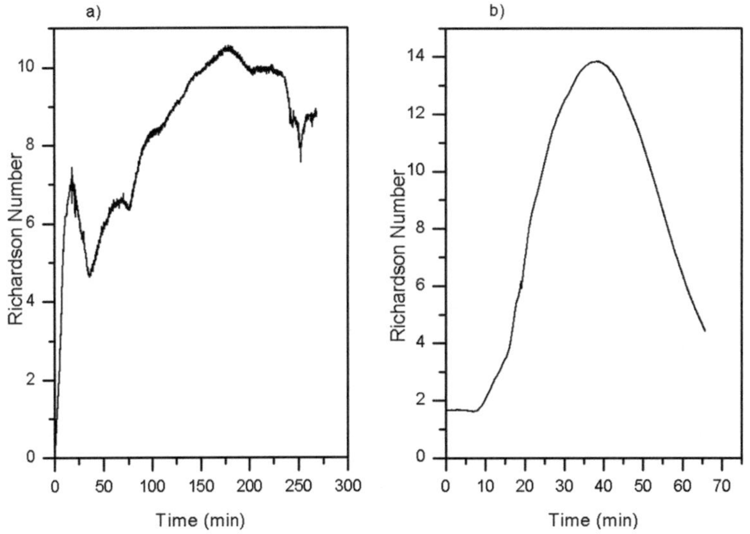

Figure 13.
The variation of the Richardson number in the storage tank during the charging and discharging process.

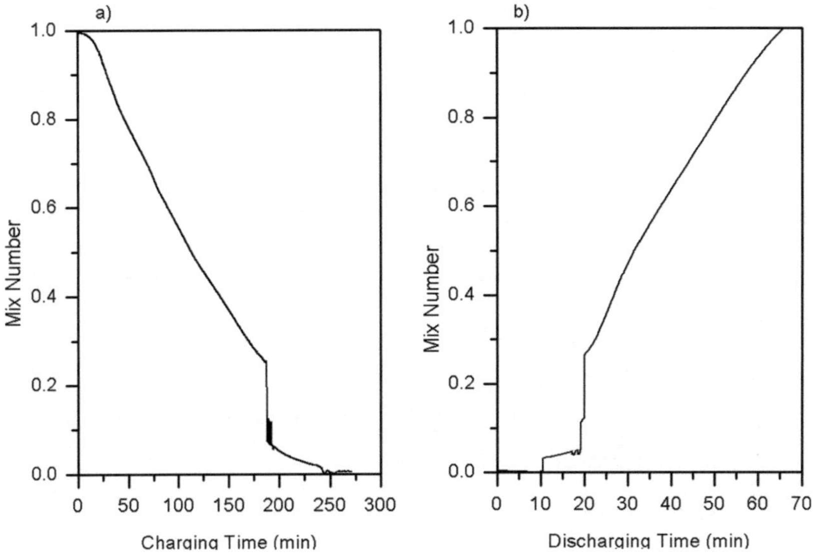

Figure 14.
Variation of MIX number during the a) charging process and the b) discharging process.

layers of HTF and later on stabilises once the temperature of HTF is around the melting point of PCM 89°C.

The stratification behaviour during the discharging process shows a steep increase after 10 minutes and stratification number reaches its maximum of 5.58 till 38 min. Then it decreases slowly and reaches a level around 2 after a period of 63 min. This infers that the maximum amount of heat can be discharged upto 40 min due to better stratification and also though stratification is reduced but significant amount of heat can be retrieved in this HEM.

Figure 13 shows the variation of the Richardson number in the storage tank during the charging and discharging process. Richardson number is a more accurate method to calculate the stratification as it includes the buoyant forces also. When the Richardson number increases it shows greater stratification whereas when the Richardson number decreases, it represents mixing forces dominate. Continuous rise in the Richardson number shows stratification found till the end and while discharging also it supports the retrieval of heat by rising gradually and then decreasing gradually after 40 minutes. The gradual increase and decrease of stratification give us an idea as to how the increase of heat charging can effectively retain and discharge the heat effectively.

Figure 14 illustrates the MIX number variation while charging and discharging. The stratification remains long as the MIX number ends reduce from 1 to 0 while charging. The MIX = 1 represents a completely mixed condition and reaches MIX = 0 for a totally unmixed condition due to charging. The reduction of MIX number is gradual due to gradual increase in stratification.

On discharging, it is observed that there is a slow and steady increase and the mix number only increases after 10 minutes so the stratification initiates. The gradual rise in MIX number shows that there is a gradual increase in stratification and represents consistent discharging of uniform heat which is very vital for any discharging application, to achieve maximum heat retrieval.

5. Conclusions

Understanding the heat exchanger design in a thermal storage helps to design a module better. To enhance this design various parameters can be utilised as discussed above. A heat exchanger with the results is being analysed and it can conclude that the amount of heat charged is easily discharged when the temperature of melting is 89°C. The average charging and discharging efficiencies are maintained high for the HEM tank as the PCM balls present retain high heat capacity and also help to maintain better stratification along with high Richardson number.

Nomenclature

Cp	specific heat of the solidus PCM, *J kg–1 K–1*
E	Energy level of the storage tank *J*
g	acceleration due to gravity, $m\ s^{-2}$
L_{pcm}	Latent heat of fusion of the PCM, $J\ kg^{-1}$
m_{pcm}	mass of the PCM, *kg*
TL1 to TL5	Temperatures of all the layers assigned in the storage tank, *K*
N	Number of temperature measurement nodes
Q	heat absorbed by the PCM, *W*
Ri	Richardson number
Str	Stratification number
$\frac{\partial T}{\partial y}$	temperature gradient along the height of the storage tank, $K\ m^{-1}$
t	instantaneous time, *s*
T	temperature, *K*
$\dot{V}V$	flow rate of the HTF, $m^3\ s^{-1}$

$\nu_{sf}v_{sf}$	superficial velocity, $m\ s^{-1}$
ΔZ	distance between the top and the bottom layers, m

Greek symbols

β	Coefficient of thermal expansion K^{-1}
ρ	density, $kg\ m^{-3}$

Subscript

avg	average
ch	charging
C_{-in}	Cold incoming water while discharging
C_{-out}	outgoing hot water while discharging
H_{-in}	incoming water while charging
H_{-out}	outgoing water while charging
i	level of the layer
Q_i	Instataneous heat loss
Q_c	Cummulative heat
Q_l	Heat loss
Q_s	Heat stored
ini	initial
k	temperature measurement nodal point counter
Lb	Bottom layer temperature
sf	superficial

Abbreviations

HEM	Heat exchanger module
HTF	Heat transfer fluid
LHTES	Latent heat thermal energy storage
PCM	Phase change material
SHTES	Sensible heat thermal energy storage
SHE	Screw heat exchanger
STHE	Shell and tube heat exchanger
TES	Thermal energy storage
CSP	Concentrating Solar Power

Author details

Mahboob E. Afshan[1*] and Anna Gowsalya Lucas[2]

1 Department of Mechanical Engineering, B.S. Abdur Rahman Institute of Science and Technology, Chennai, Tamil Nadu, India

2 Department of Production Technology, Madras Institute of Technology, Anna University, Chennai, Tamil Nadu, India

*Address all correspondence to: mahboobemech@gmail.com

References

[1] Sarbu I, Dorca A. Review on heat transfer analysis in thermal energy storage using latent heat storage systems and phase change materials. International Journal of Energy Research. 2019;**43**:29-64

[2] Dincer I, Rosen MA. Thermal Energy Storage: Systems and Applications. 2nd ed. London: Wiley

[3] Haller MY, Cruickshank CA, Streicher W, Harrison SJ, Andersen E, Furbo S. Methods to determine stratification efficiency of thermal energy storage processes – Review and theoretical comparison. Solar Energy. 2009;**83**:1847-1860

[4] Cabeza LF. Advances in Thermal Energy Storage Systems: Methods and Applications. Woodland Publishing, Elsevier; 2015

[5] Sarbu I, Sebarchievici C. A comprehensive review of thermal energy storage. Sustainability. 2018;**10**:191

[6] Oro E, Albert C, Chiu J, Martin V, Cabeza LF. Stratification analysis in packed bed thermal energy storage systems. Applied Energy. 2013;**109**: 476-487

[7] Cabeza LF, Ibanez M, Sole C, Roca J, Nogues M. Experimentation with a water tank including a PCM module. Solar Energy Materials. 2006;**90**: 1273-1282

[8] Afshan ME, Selvakumar AS, Velraj R, Rajaraman R. Effects of aspect ratio and dispersed PCM balls on the charging performance of a latent heat thermal storage unit for the Solar thermal applications. Renewable Energy. 2020; **148**:876-888

[9] Afshan ME. Experimental Investigation on the Charging and Discharging Performance of a Thermal Storage Unit Due to Stratification Affected by Aspect Ratio and Dispersed PCM Balls. Ph.D thesis. 2020

Chapter 3

Techniques for the Thermal Analysis of PCM

Abdelwaheb Trigui

Abstract

Thermal Energy Storage (TES) technologies based on Phase Change Materials (PCMs) with small temperature differences have effectively promoted the development of clean and renewable energy. Today, accurate thermal characterization is needed to be able to create an optimal design for latent heat storage systems. The thermo-physical properties of PCMs, namely latent heat, phase-change temperatures, enthalpy and specific heat capacity are obtained by means of differential scanning calorimetry (DSC), which is one of the most widely used techniques to study reactions related to the transformation of a material subjected to temperature constraints. This method presents some limitations due, among other things, to the fact that only a very small quantity (less than 90 mg) of material can be tested. Indeed, the small mass samples, taken out of the large testing specimen and out of testing system, is not representative of the thermal behavior of a material on a large scale. The Transient Guarded Hot Plate Technique (TGHPT) presents several advantages when compared to the commercially available thermal analysis methods (DSC, DTA) to determine PCM thermophysical properties. The most significant are large sample amount, optimized measuring time and a simple and economical built up.

Keywords: thermal energy storage (TES), phase change materials (PCMs), differential scanning calorimetry (DSC), transient guarded hot plate technique (TGHPT)

1. Introduction

Phase change materials (PCMs) are thermal storage materials with a high storage density for small temperature range applications. The determination of the thermophysical properties is a key step not just for the application itself but also for the material selection to define the suitability of a material for use in TES. The choice of the suitable thermal analysis method for experimental data acquisition depends on the device outputs, measured values accuracy, experimental setup requirements as sample size, influence of heating/cooling rate maintenance and equipment price, the implementation, etc. There are different thermal analysis techniques to characterize materials and the convenience to use them relies on the properties that want to be determined. The non-steady-state or transient technique records a measurement during the heating/cooling process. The method determines the enthalpy change of the PCM as a function of temperature with high precision by means of transient sensors. Depending on the application, the amount of PCM used can vary from a few (milli) grams (e.g., cooling of electronic equipment or clothes) up the kilograms (e.g., storage in a solar thermal power plant or in the walls of buildings in thermal management and occupant comfort). To determine the specific heat and latent heat of materials a number of thermal characterization techniques help to provide better results especially differential scanning calorimetry (DSC) are commonly used [1]. DSC is an effective method to characterize the thermal behavior of PCMs, and to determine their TES capacities, in terms of transition temperature, latent heat and specific heat capacity and its stability throughout the various melting and crystallization cycles. Using the data measured by the DSC method, it is also possible to represent the enthalpy change versus temperature and determine the amount of stored/released energy in a given temperature interval. In case of the heat capacity measurement for a sample which does not undergo phase change, the energy supplied is weak and generally not very variable. On the other hand, in the case of a fusion process, there is a rapid transient which require important heat rates from the DSC. The thermal imbalance between the two cells is then very important although the quantity of product remains low. DSC presents some problems in analyzing PCM due to their high enthalpy density per unit volume: the small sample amount (less than 90 mg), the sample size influence on its thermal behavior, the response dependence on the used heating rate, repeatability when undergoing huge number for solidification/melting cycle. For large size samples, the melting process occurs gradually through the material. The latter is then heterogeneous and the two phases may coexist over long periods of time before a complete fusion. Moreover, heat conduction in the solid and convection in the liquid occur [2]. This strongly influences the global (or apparent) behavior of the PCM. In practical application, the material volume is much more important, up the kilograms for instance, so testing using a noninvasive method for latent and specific heat determination is necessary. It was found that identification of thermophysical properties of PCMs over several cycles (solidification and fusion) requires the design of a new experimental device Transient Guarded Hot Plate Technique (TGHPT). The TGHPT being an alternative to DSC measurements allows one to determine the same thermophysical properties as DSC.

2. Differential scanning calorimetry (DSC)

DSC is nowadays the most used technique to determine the melting/solidification points and the latent heat of phase change and specific heat capacity of PCM. It is also useful to observe other phenomena such as supercooling, hysteresis and glass transition [1]. The TA Instruments DSC is a "HEAT FLUX" type system where the differential heat flux between a reference (e.g., sealed empty Aluminum pan) and a sample (encapsulated in a similar pan) is measured. The measuring device operates in the temperature range from −150 to 700°C. To attain information on the heat flows corresponding to the temperature fluctuations the differential scanning calorimeter uses an operation analogue to Ohm's Law. In DSC analysis, the equipment has two melting crucibles, one empty and used as the reference, and the other with the substance to analyze (sample). Then it is programmed an ascendant ramp of temperature versus time. It was found that the difference between the two melting pots to obtain a signal. A schematic diagram of Heat Flux DSC is shown in **Figure 1**.

The sample and the reference are placed symmetrically in the furnace. The furnace is controlled under a temperature program and the temperature of the sample and the

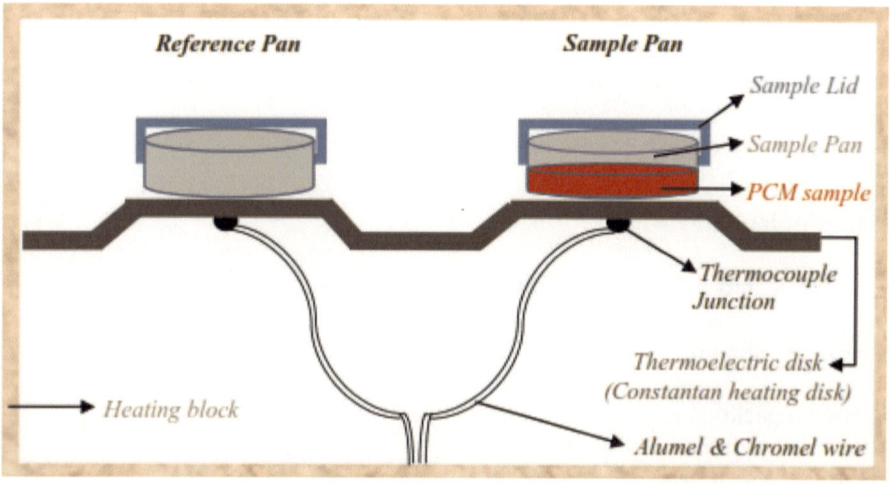

Figure 1.
Schematic diagram of Heat Flux DSC.

reference are changed. During this process, three situations can occur with respect to the sample:

- no reaction or phase transition and no peak results
- The PCM thermal properties (melting temperature, latent heat) during the endothermic peak results
- The thermal measurement data (solidification temperature, latent heat) during the exothermic phase transition process.

The thermogram is a plot of heat flow against temperature profile. During thermal process reactions either liberate or absorb heat. Thus, when ΔH is positive (endothermic reaction), the sample heating device is energized and a positive signal is obtained; when ΔH is negative the reference heating device is energized and a negative signal is obtained. The peak areas in DSC are proportional to the amount of sample, the heat of reaction.

Commonly, calorimetric measurements on PCMs affect the shape, precision and accuracy of the experimental results are:

2.1 Instrumental factors

- Furnace heating rate (A fast heating rate may minimize the data acquisition time compromising salient features of the material property)
- Obtain a thermal profile with a calibration standard (e.g., sapphire, water, gallium and indium) using the same scan conditions. Provides a calibration factor that translates the measured Heat Flow units in (mW) to Heat Capacity units (J/g°C).
- Furnace atmosphere (e.g., inert, argon, nitrogen, oxygen, helium ...).

- Geometry of sample holder and location of sensors (variations in baseline).
- Sensitivity of recording mechanism.
- Composition of sample container.

2.2 Sample characteristics

- Structure and amount of the sample (proper thermal contact).
- Heat of reaction.
- Sample packing.
- Nature of sample.
- Thermal conductivity.

2.3 Characterization PCM materials using DSC techniques

DSC technique allows to quantitatively determine in situ the thermodynamic changes occurring during the transformation of a solid into a liquid (heating) which directly characterizes PCM thermal properties (i.e., enthalpy, specific heat capacities and melting point).

For a TES system, it is necessary to take into account some characteristics that a phase change material must meet in order to be considered competitive and suitable for its function. These are physical, chemical, thermodynamic, kinetic and technical characteristics [3]:

- Phase change temperature suitable for the process or application.
- Phase change taking place at a unique temperature, or in a very small phase change interval.
- High phase-change enthalpy to provide a high energy storage density.
- Low price, to be competitive with other thermal storage options (heat or cold).

The correct choice of DSC measurements conditions is very important input to thermal characterize of PCM to obtain phase change temperature range and the relationship of specific heat capacity with temperature (Cp-T). Accurate determination of phase transition parameters should provide robust insights about the performance of the PCM, including energy storage capacity and phase change temperature, during several charging and discharging cycles, to meet the design requirements of thermal energy storage applications and to prevent potential failures. Based on the typical DSC thermogram, shown in **Figure 2**, the measured melting temperature range was usually determined manually. While some researchers recognized the temperature range between the onset temperature (T_{Onset}) and the endset temperature (T_{endset}) as the actual melting temperature range. The endset melting

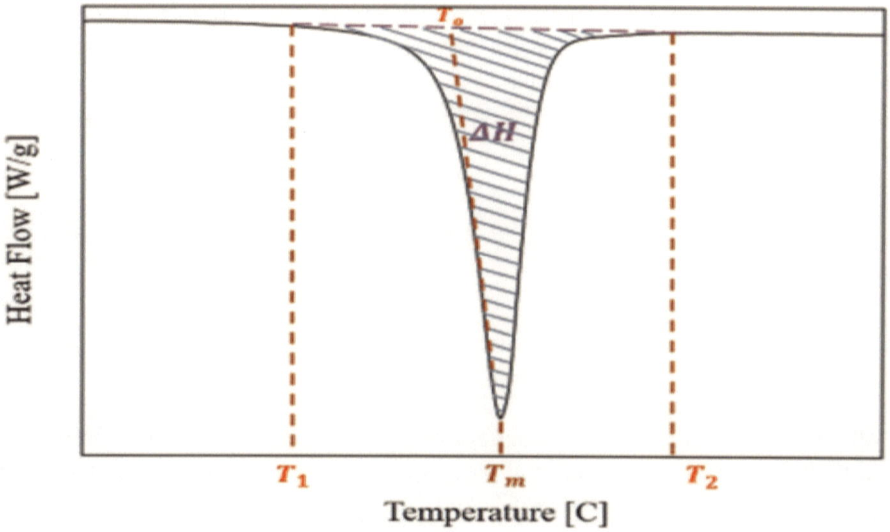

Figure 2.
Typical characteristics of a DSC thermogram. The blue shaded area (ΔH) represents the enthalpy change during transition (total latent heat) of PCM, T_1 and T_2 are the temperature range at which the transition process occurs, T_m: Melting temperature, and T_{onset}: Is the onset temperature.

temperature depends on the measurement conditions (heating rate, sample mass, heat transfer).

From the sample heat–flux signal, the enthalpy $h(T)$ is determined by integration of every peak. The enthalpy-temperature curves are more useful in PCM research field and can be determined as the sum of the enthalpy intervals. Compared to the DSC thermogram, the enthalpy-sum curves perform the best representation to determine the sum of both latent and sensible heat as a function of temperature. The current way has gained increasing attention to characterize and compare the materials by one curve only.

It is of vital significance the determination of the enthalpy of PCM as a function of temperature with sufficient accuracy in enthalpy and phase change temperature range is an important data especially for the the numerical analysis of TES system.

Based on the recorded heat flow (DSC signal), changes in enthalpy (latent heat) or the specific heat capacity (sensible heat) of an examined sample are determined by recording the absorbed heat between two equilibrium states, assigned as baselines of the acquired measurement curves [4]. From the specific heat, the heat flux can be obtained as a function of temperature, and the specific enthalpy is determined easily by integration procedures from Eq. (1), as follows:

$$\Delta H = \int_{T_1}^{T_2} C_p\ (T).dT = \int_{T_1}^{T_2} \frac{\partial Q/\partial t}{dT/dt}.dT = \int_{T_1}^{T_2} \frac{\partial Q}{\partial T}.dT \tag{1}$$

Where Δh is the enthalpy (latent heat or energy storage between the temperature increments T_1 and T_2) in units of J/g, Cp is the specific heat capacity at constant pressure in units of J/(g.K), T_1 and T_2 represent the temperature range at which the storage operates, $\partial Q/\partial t$ is the DSC heat flow signal in units of W/g, and dT/dt is the DSC heating rate in units of °C/sec.

As shown in **Figure 3**, the melting latent heat (ΔH), total energy storage capacity, melting temperature (T_m), and specific heat capacity of the liquid and solid phases can be directly taken from the curve.

The enthalpy sum as a function of temperature can be determined using Eq. (2) as follows:

$$H_{sum}(T_i) = \sum_{k=0}^{i-1} H(T_k) + H(T_i)$$

$$H(T_i) = \int_{T_i}^{T_{i+1}} \frac{\partial Q/\partial t}{dT/dt} * dT \quad (2)$$

Where $H_{sum}(T_i)$ is the enthalpy sum or energy storage between the reference's initial temperature (T_0) and the final temperature (T_i), H_{Ti} is the enthalpy at a single temperature increment between two T_i and T_{i+1}, $\partial Q/\partial t$ is the heat flow in units of *W/g*, and dT/dt is the DSC heating rate in units of °C/sec.

For accurate results, changes in enthalpy (latent heat) or the specific heat capacity (sensible heat) of an examined sample are determined by recording the absorbed heat between two equilibrium states, assigned as baselines of the acquired measurement curves. It is worth highlighting that the baseline-construction due to the DSC measurement of PCMs requires a careful procedure to achieve reliable results.

1. The heating/cooling rate effects (Trade-off between sample attaining thermal equilibrium and data acquisition times. This is done by using the real sample measurement for each temperature range and an accurate calibration using the heating rate which will be applied to the sample starting from fast rates and slowing down the heating and cooling rates of consecutive cycles by halves the previous).

2. Calibration of sensors is only one application of reference materials (e.g., sapphire, water, gallium and indium). An accurate calibration using the heating rate which will be applied to the sample is crucial with regard to all these usages

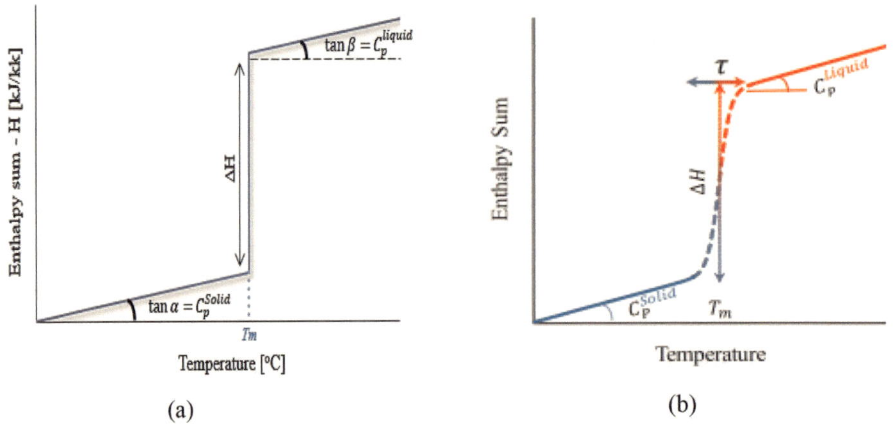

Figure 3.
The enthalpy sum as a function of temperature for (a) an ideal PCM, (b) real PCM.

thus must cover not only the temperature range, but also specifics of equipment including sample size, cost, and safety etc.

3. Measurement of the empty crucible using the determined heating and cooling rates.

4. Structure and mass of the sample (proper thermal contact) using the selection of the proper "heating rate".

    a. Each sample test was repeated four times to ensure reproducibility of the experiments.

    b. Considering 2 different (low and high) sample masses to be tested.

5. Analysis of data

    a. Sample positioning on the DSC stage (the baseline would be a flat line of zero milliwatts).

    b. Subtract the baseline data from the sample and the calibration standard (e.g., sapphire) data, prior to analysis [Remember to change the Heat Flow (y-axis) units to mW, before subtracting the data, to ensure identical units for the sample and baseline profiles].

    c. Data evaluation; the latent heat capacity of melting and freezing of each sample was determined by numerical integration of the area under the peaks of phase change transitions.

The specific heat capacity is a key parameter for PCM, this physical property is essential to conduct sensible heat process (before and after phase transitions). From the known masses of PCM and reference, and the specific heat capacity of the standard, the specific heat capacity of the PCM is calculated based on the DSC heat-flow curve at any point of time. The conversion between time and temperature is finally done via the recorded temperature ramp.

The specific heat capacity of the PCM material can be expressed using Eq. (3):

$$C_p(T) = \left[\frac{60\ E}{H_r}\right] * \frac{\Delta y}{m} \tag{3}$$

Where $Cp(T)$. is the specific heat capacity of the PCM material at the temperature (T) of interest, E: cell calibration coefficient at the temperature of interest (dimensionless), H_r: the heating rate in °C/min, and Δy : the deflection in the y axis between the baseline curve and sample measurement curve at the temperature of interest in mW and m is the sample mass in mg

As seen in **Figure 3**, heat capacities and changes in heat capacity can also be determined from the enthalpy-temperature curves. While changing phase during a measurement, the sample is far away from thermal equilibrium. In contrast to materials without a phase change or with high thermal conductivity, the slope of the sensible heat part (the solid and liquid phase region) in the enthalpy-temperature curve is equal to the specific heat capacity of the tested material.

The heat capacity of the sample is computed by assuming that the thermal resistance of reference and sample crucible is the same. The thermal resistance is determined by a calibration using standard materials with well-known heat of fusion. The temperature sensors by the melting temperatures (onset temperatures) of these materials. In general, for a solid to liquid phase change process, the specific heat at a constant pressure is the energy stored when it experiences a temperature change of 1°C. The heat capacity can be found in the Eq. (4) as follows:

$$C_p(T) = \frac{\partial Q}{\partial T} \tag{4}$$

In the case of ideal PCM, the heat storage occurs over three distinct thermal events. When a PCM at solid state is heated and its temperature is raised uniformly from the system starting temperature until it reaches the phase change point. Over this period, heat is stored in sensible form. Further heating of the material incurs solid to liquid phase transition and heat is stored in latent form and temperature is constant during this period. Once the phase change is completed, the PCM is in liquid phase, further heating of the material increases its temperature and heat is stored again in sensible form in a rate proportional to the PCM specific heat. The reverse of the processes described above releases heat, which is also called discharge process. As can be seen, the total amount of energy stored released for a TES system involves three stages, two for sensible storage and one for latent storage, and hence the total heat storage capacity of a LHTES material can be calculated by Eq. (5) [4]:

$$Q = m\left[\int_{T_1}^{T_m} C_{p-solid}.dT + \int_{T_m}^{T_2} C_{p-liquid}.dT\right] \tag{5}$$

Where $C_{p-solid}$ is the average specific heat between $T_1$and T_m, $C_{p-liquid}$ is the average specific heat between T_m and T_2, m: mass of PCM, T_1 and T_2 are the temperature range in which the TES process operates, T_m: melting temperature, and Δh: latent heat of solid-liquid transition.

The total energy stored "Enthalpy-sum" for PCMs can be calculated according to the upper and lower temperature limits of the TES system, as shown in Eq. (6):

$$E_{stored} = m_{pcm} * [H(T_2) - H(T_1)] \tag{6}$$

Where E_{stored} is the total energy stored in the PCM TES system, m_{pcm}: the total mass of PCMs in the TES system, $H(T_2)$: the enthalpy-sum at the upper temperature limit (T_2) of the TES system, and $H(T_1)$: the enthalpy-sum at the lower temperature limit (T_1) of the TES system. As clearly seen in **Figure 4**, as considered in the enthalpy-sum curves, the heat capacity of a LHTES material depends on its specific heat of both solid and liquid and the latent heat. Therefore, a high heat capacity and a large latent heat are key for LHTES materials selection. As mentioned above, the transformation between solid and liquid states is specifically attractive due to small change in volume and has therefore been mostly investigated and utilized [5].

Case studies will be discussed to demonstrate many of the capabilities of this sensitive analytical instrument heat flux DSC. The PCM (RT27) as an example was studied according to the following experimental procedure [6]:

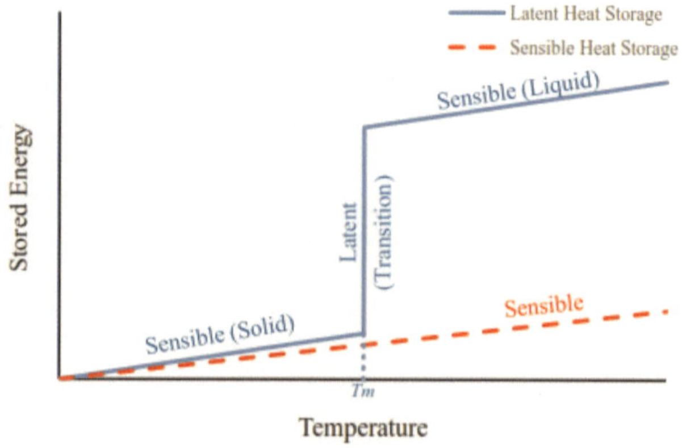

Figure 4.
Performance comparison of the storage mechanism for TES.

1. Cooling from 40 to −20°C at 10°C/min
2. Heats from −20 to 40°C at 10°C/min
3. Cooling from 40 to −20°C at 5°C/min
4. Heats from −20 to 50°C at 5°C/min
5. Cooling from 50 to −60°C at 0.2°C/min
6. Heats from −60 to 50°C at 0.2°C/min

Trigui et al. [6] reports a comparison of thermophysical properties of paraffin wax (RT27) obtained for two samples as summarized in **Table 1**. When using a DSC, the heat flux sensor is calibrated and the measurement procedure depends on the calibration mode, which is heat capacity or enthalpy calibration. With the using the enthalpy calibration, only the peak integral can be determined. That means that the sensible heat is not taken into account, and that there is no temperature resolution within the peak [7]. Using the heat capacity calibration, it is possible to determine the sum of latent and sensible heat as a function of temperature.

2.4 Problems related with PCM analysis using DSC

In the selection and analyzing *PCM* due to their high enthalpy density per unit volume, the most problematic aspects are generally sample size, the stability to thermal cycling, phase separation, large supercooling and poor thermal conductivity. Moreover, DSC test results depend on further factors such as sample preparation, correct calibration to improve the accuracy of the PCM characterization procedure and to make measurement errors negligible [8].

- *Sample size:* The heating and cooling rates of DSC measurements are typically much faster than in real applications, while analysis on very small sample size of PCM (typically 20μL for the DSC analysis) is not completely representative of the

Properties		**Values measured**	
		DSC calibrated on *Cp*	**DSC calibrated on *ΔH***
		Sample1 {**Mass, 12.47mg**; **Scanning rate, 10°C/ min**}	**Sample2** {**Mass, 5.11mg**; **Scanning rate, 0.2°C/ min**}
Heating state	solid-solid phase change temperature, T_{pm}	5.08°C	3.58°C
	Melting point, T_m	25.53–25.69°C(*)	23.67–23.70°C(*)
	Latent heat of solid-solid change, L_{pm}	23.692 J/g	23.300 J/g
	Latent heat of melting, L_m	155.960–140.581 J/g(*)	134.928–158.110 J/g(*)
Cooling state	solid-solid phase change temperature, T_{pc}	3.95°C	3.23°C
	crystallization point, T_c	26.2°C	24.78°C
	Latent heat of solid-solid change, L_{pc}	22.594 J/g	22.031 J/g
	Latent heat of crystallization, L_c	162.165–154.848 J/g(*)	130.139–161.087 J/g(*)
Specific heat of solid, $C_{p,s}$ (≤15°C)		3.25 ± 0.35 J/g.°C	
Specific heat of liquid, $C_{p,l}$ (≥40°C)		2.23 ± 0.005 J/g.°C	

(*) Value measured by sigmoïde.

Table 1.
Thermophysical properties of paraffin wax (RT27) [6].

thermal properties of the PCM bulk material, which might lack representativity for real size applications [9–11].

The sample size also affects the signal of the sample. If a small sample size is selected with low heating and cooling rates, the temperature shift inside the sample will be reduced. But this is not the solution. Because, both the small sample size and low heating and cooling rate lead to a weak signal and hence, decreasing the accuracy in enthalpy.

- *Supercooling:* Supercooling, also called undercooling, is often stronger for small sample sizes than in large sample sizes. Supercooling is the effect under which many PCM do not solidify immediately upon cooling below the melting temperature, but start crystallization only after a temperature well below the melting temperature is reached. Therefore the sample size should be large enough to obtain the real behavior of the sample. If the sample present strong supercooling in DSC analysis, it deforms the cooling DSC curve. Due to the supercooling, it becomes difficult to quantify internal gradients by comparing heating and cooling curves since the latent heat is released at a lower temperature than the intended one, and part of this heat is used to increase the temperature of the material to the melting temperature, where final crystallization occurs. Hence, as a solution to the supercooling, the large sample size should be analyzed before applying the PCM for any real TES system.

- *Calibration:* For calibration, materials with a known thermal effect are used. When using a DSC, the heat flux sensor has to be calibrated and the measurement procedure depends on the calibration mode, which is heat capacity or enthalpy calibration. Two common calibration modes are heat capacity (also called heat flow rate) calibration and enthalpy (also called heat) calibration. Each of the two calibration modes has advantages but also disadvantages with respect to PCM as sample. Using the heat capacity calibration, it is possible to determine the sum of latent and sensible heat as a function of temperature. Another approach to calibration is using standard materials with a well known melting enthalpy [9–11].

The thermal resistance is done by comparing measured phase change temperatures of standard materials with well-known heat of fusion and the temperature sensors by the melting temperatures (onset temperatures) of these materials.

- *Thermal equilibrium:* For reaching thermal equilibrium the sample should be isothermal, otherwise thus producing a shifting of the phase change temperature and also lead to non-realistic shapes of the heat capacity and/or enthalpy temperature responses $h(T)$. If the measured temperature should be close to the real sample temperature, the heating or cooling should be slow to diminish the temperature gradient between sample and crucible. Therefore, the thermal equilibrium status in the sample is essential to improve the accuracy of the PCM characterization procedure and to make measurement errors negligible [9–15].

- *Hysteresis:* There are mainly two reasons for the hysteresis phenomenon of PCM. One is the improper measurement method can lead to the observation of an "apparent hysteresis," which in most cases, over-estimates the real hysteresis of the material. Continuous heating and cooling can also increase the apparent hysteresis. The other reason for the hysteresis phenomenon of PCM lies in its intrinsic material property. Decreasing the heating and cooling rates, sample size or temperature step can improve the measurement accuracy so that the measurement uncertainty decreases. However, the single enthalpy curve without hysteresis cannot be achieved even with infinitely slow measurement, because the existence of subcooling, incomplete crystallization or polymorphic crystal structures results in different shapes of heating and cooling curves [16–18]. In certain cases, even with the most accurate measurement method and with the smallest heating/cooling rates, the hysteresis phenomenon can still be observed.

3. Transient guarded hot plate technique (TGHPT)

The Transient Guarded Hot Plate Technique (TGHPT) provides an alternative approach to characterize some of the thermo-physical of PCMs. Most of thermal analysis techniques, such as the DSC method are designed to test small, pure and homogeneous samples. During long-term performance testing, when undergoing huge number for freezing/melting cycles, large PCM specimens often are subject to settling or stratification and are typically not perfectly homogeneous. The TGHPT being an alternative to DSC measurements allows one to determine the phase change material (PCM) thermal properties (thermal conductivity λ, sensible and latent heat thermal energy storage, C_P and L_m latent heat of fusion or melting enthalpy) in the solid phase, during the solid–liquid transition and in the liquid phase and its thermo-chemical stability after thermal cycling test. The TGHPT eliminates the previously mentioned limitations of DSC measurements (i.e., mass influence, low thermal conductivity and thermal diffusivity of PCMs, non-equilibrium thermal gradients in the DSC pan and heating/cooling rate) and also allows for measurement of much larger samples than DSC (about 1000 times larger) [19]. The TGHPT has some advantages over other methods: it can be used, due to large sample size, for a wide variety of PCM (inorganic and organic, being encapsulated or composite), heating and cooling rates and temperatures ranges are variable and large enough to fit PCM for different applications. Therefore, for PCM characterization large samples are preferable. The TGHPT method has also the advantages of (i) a simple experimental device as it can be built up with basic laboratory devices (ii) lowest maintenance and lowest price. To clarify the effects on the heat transfer performance, an experimental device allows to determinate changes in enthalpy (latent heat) or the specific heat capacity (sensible heat), thermo-physical properties like thermal conductivity of an examined sample by Trigui et al. [6] (**Figure 5**). The proposed test bench for the parallelepiped-shape of composite (4, 5 × 4, 5 × 0.6 cm^3) allows the simultaneous measurements of the temperature and heat flux at the material edges. The PCM material is placed between two horizontal exchanging aluminum plates. The plates get tempered by fluid thermostats with a precision of about 0.1°C. Each side of composite is equipped with a heat flux sensors and thermocouples (type *T*). These heat flux sensors fixed at the sample surfaces to measure input and output heat flux have 0.2 mm of thickness, 202 $\mu V/W/m^2$ of sensitivity and 400 cm^2 surface areas. The sensors were connected to LabVIEW software to measure temperature fluctuations and heat flux exchanged during melting and solidification processes which allow the investigation of the thermal stability of the samples. The acquisition of the experimental data is recorded at regular and adjustable time steps (1–6 s). To ensure a one dimensional heat flow through the sample the lateral sides are insulated by 11 mm thickness of expanded polystyrene. The enthalpy–temperature curve of a sample is measured similar to DSC measurements by heating up and cooling two aluminum heat exchanger plates in

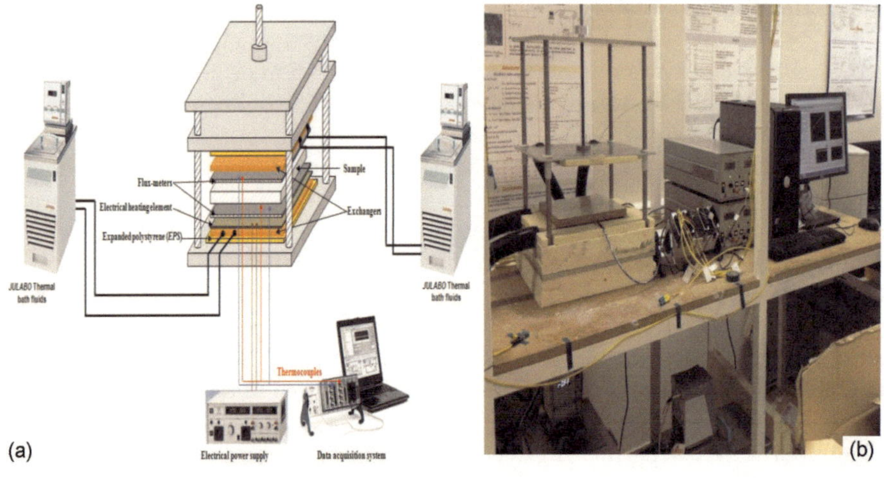

Figure 5.
(a) Scheme of the Transient Guarded Hot Plate Technique (TGHPT) for testing PCM materials, (b) Picture of the whole apparatus.

parallel at a certain heating rate or in stepwise mode. The TGHPT was developed to test construction materials with PCM in large scale samples in different boundary conditions following:

- A steady-state temperature difference on both sides of the sample imposed by two exchanger plates to determine the thermal conductivity.
- A transient regime between two stationary regimes to measure the variation of the specific heat versus temperature and interval where the two phases coexist to measure latent heat of melting.

With the TGHPT, the thermal performance of PCM such as the thermal conductivity, its temperature in liquid and solid state for PCM, the specific heat capacity and the latent heat can be determined. Furthermore, it is possible to investigate the thermal performance of PCM material such as the thermal stability and repeatability to store and release the thermal energy.

The TGHPT optimize measurement with shorter experimental time than DSC.

3.1 Apparent thermal conductivity

The experimental sample is placed in between two aluminum heat exchanger plates, and the position of the plates is adjusted to establish a full contact with the surface of the measured sample. To determine experimentally solid and liquid thermal apparent conductivities of the composite, the apparatus establishes a steady state one-dimensional heat flux through the test sample between the two plates. The plates are set a at constant but different temperatures to establish a thermal gradient through the sample. The temperature difference between the top and bottom plates is a user defined value to perform measurements at various averaged temperatures. The apparent thermal conductivity can be calculated using Fourier's law of heat conduction. The thermal conductivity of the test sample is given by Eq. (7) [3, 20, 21]:

$$\lambda_{s,l} = \frac{e.\sum \varnothing_{s,l}}{2.\Delta T_{s,l}} \tag{7}$$

where e is the thickness of the specimen; $\Sigma\varphi_{s,l}$ is the sum of the measured heat fluxes of the solid or liquid states.

As an example, the composites materials (LDPE/Wax) were studied for thermal energy storage with a melting point around 26°C [6]. In order to characterize the apparent thermal conductivity of the solid phase, the TGHPT has been used with temperature variation only in a single face of the sample. The state of initial balance ($T_{init} = 15°C$, lower than the melting temperature (T_m)) is brought back towards another state of final balance ($T_{end} = 20°C$), such that $T_{end} < T_m$) (fusion process)) where heat flux tightens towards a non null value corresponding to a temperature gradient generated by the two heat exchangers plates. For the liquid phase, the same experimental procedure was applied by choosing a temperature difference (40 and 50°C). **Figure 6** summarizes the measured thermal conductivities and associated uncertainties of the composites with different Wax contents.

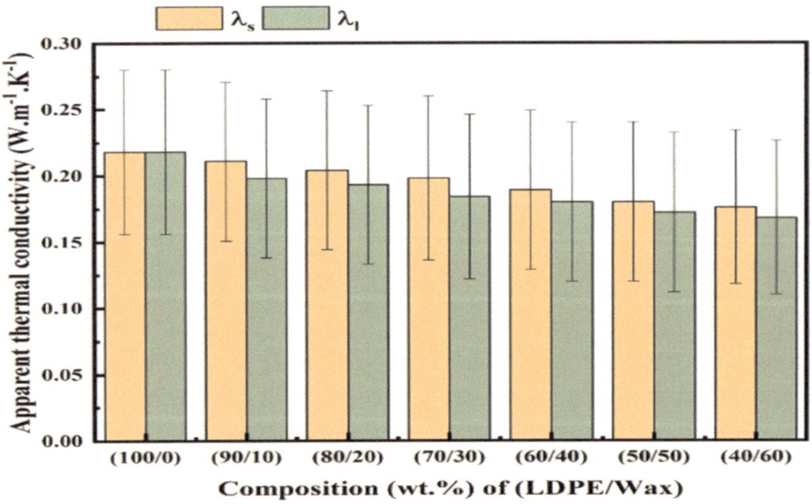

Figure 6.
Thermal conductivities of LDPE/Wax composites [6].

3.2 Sensible heat and apparent heat capacity (in solid and in liquid states)

At the beginning, the two aluminum heat exchanger plates and the composite were maintained at a constant temperature T_{init}. By modifying the temperature set point of the thermo regulated bath, the composite evolved to a steady isothermal final temperatureT_{end}. Between both states, the PCM composite stored an amount of sensible heat, which represents the internal variation of the system's energy [22]. The heat capacity of the material was determined for the solid and liquid phases by calculating the integral of the heat flux difference from the initial state (t_{init}) and the final state (t_{end}) using the following expression Eq. (8) [22, 23].

$$Q_{sens} = \frac{1}{\rho\, e} \int_{tinit}^{tend} \Delta\varnothing.dt = C_p.\ (T_{end} - T_{init})\ [kJ/kg] \tag{8}$$

Where C_P is the apparent specific heat capacity of the composite, $\Delta\varnothing$ represents the difference heat flux measured at each time step during the acquisition dt, e is the composite's width and ρ is its density.

To evaluate the apparent specific heat of the PCM, Different tests were carried out:

1. The temperature was checked from $T_{init} = 15°C$ to $T_{end} = 20°C$ to determine the sensible and specific heats of the compound in the liquid state.

2. The temperature was checked from $T_{init} = 40°C$ to $T_{end} = 50°C$ to determine the sensible and specific heats of the compound in the solid state.

Figure 7 presents the evolution of the heat fluxes and temperatures on both sides of two samples during the determination of the specific heat capacities of the LDPE/Wax composite in the solid and liquid states respectively. For the liquid and solid phases, the measured temperatures on the lower (T_1) and the upper (T_2) faces of the material evolve in an asymptotic way to the set point. Also, we can note that the flow

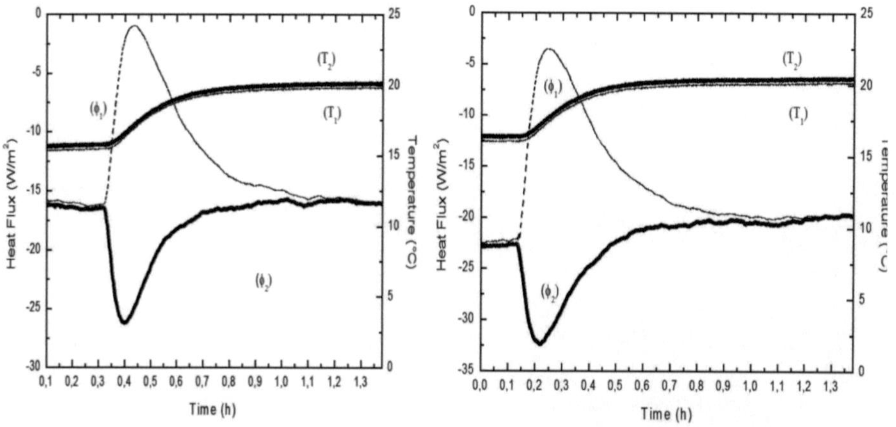

(a) LDPE/wax (90/10) *(b) LDPE/wax (40/60)*

Figure 7.
Heat flux and temperatures evolution of the solid phase (15–20°C) [6].

Samples	Q_{sens} (kJ /kg)		Cp (kJ /kg. °C)		Q(kJ/kg)	L_m (kJ/kg)
	solid (15–20°C)	liquid (40–50°C)	solid (15–20°C)	liquid (40–50°C)	evolution (15–50°C)	
LDPE/wax (90/10)	13.8	24.9	2.8	3.5	104.0	65.3
LDPE/wax (40/60)	20.6	32.1	4.1	3.2	159	106.2

Table 2.
Quantity of heat stored and apparent heat capacity for (LDPE/Wax) composites [6].

evolves very quickly at the beginning of recording and then to a steady state obtained at the end of the test. The results obtained for the specific heat capacity and the sensible heat stored for the solid and liquid states are given in **Table 2**.

3.3 Storage and release of latent heat

LHTES is the most promising method in this field due to excellent phase change behavior and high heat storage capacity. The amount of total energy storage/release can be calculated by a temperature variation from 15 to 50°C. Between these two positions, the composite stores and releases a sensible (Q_{sens}) heats and latent (L_m) heats. The Latent heat is calculated by subtracting the sensible heat from the total amount of heat. This quantity can also be expressed by Eq. (9) [24, 25]:

$$Q = Q_{sens} + L_m = (C_{ps}.\ \Delta T_s + C_{pl}.\ \Delta T_l) + L_m\ [kJ/kg] \tag{9}$$

where C_{ps} and C_{pl} are the heat capacities of composite at solid and liquid state, respectively, ΔT_s and ΔT_l are the temperature variations in solid and in liquid state and L_m is the latent heat par unit mass of melting.

The selected temperatures (15–50°C) are sufficiently far from the melting point to consider the material as being strictly in one state or the other. **Figures 8** and **9** show that the heat stored is much more important than sensible heat transfer influenced by

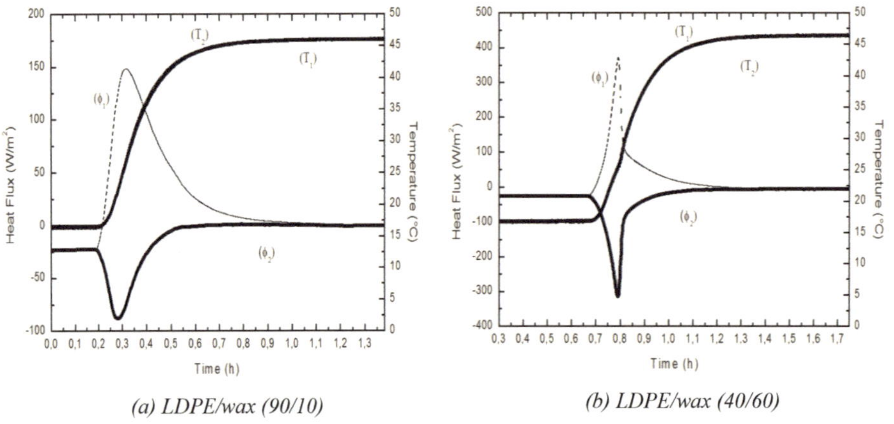

Figure 8.
Heat flux and temperatures evolution from solid to liquid (15–50°C) [6].

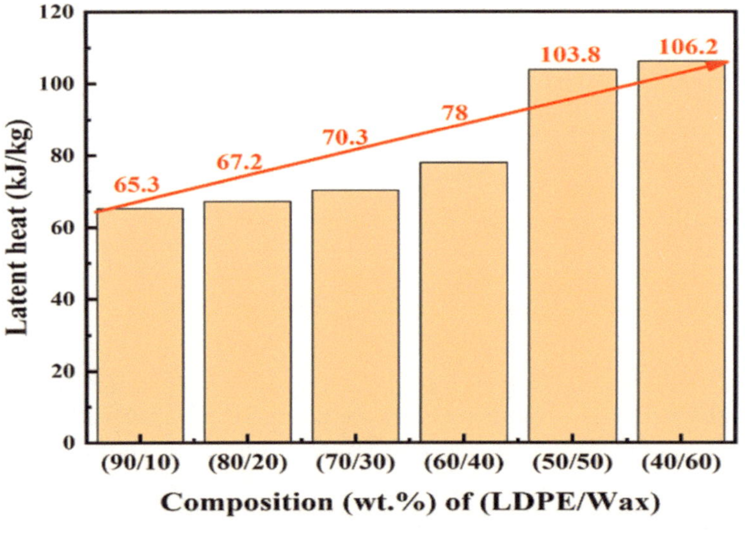

Figure 9.
Latent heat of LDPE/Wax composites [6].

the paraffin wax addition. Hence, the addition of PCM significantly improves the thermal performance of composite in terms of saving energy.

Figure 10 shows present the evolution of the heat fluxes and temperatures on both sides of the composite during the solidification process where the temperature evolves from 15–50°C. Initially, it can be observed a normal evolution of measured heat flux corresponding to the cooling of the liquid phase. At the end of this phase (t = 33 min), when the temperatures of surfaces are in the vicinity of 28°C, it can be seen the release of latent heat, and the appearance of a distinct inflection point in the heat flux curves. From this critical moment, the cooling of the sample continues, the material is solidified slowly and cools until it reaches the prescribed 15°C: after more than 1 h30 min, the sample reaches an equilibrium state. Notably, heat restitution was a very long process since at the start of the solidification. At the beginning of solidification, a layer solid Hexadecane was formed at the upper surface of the material which was in

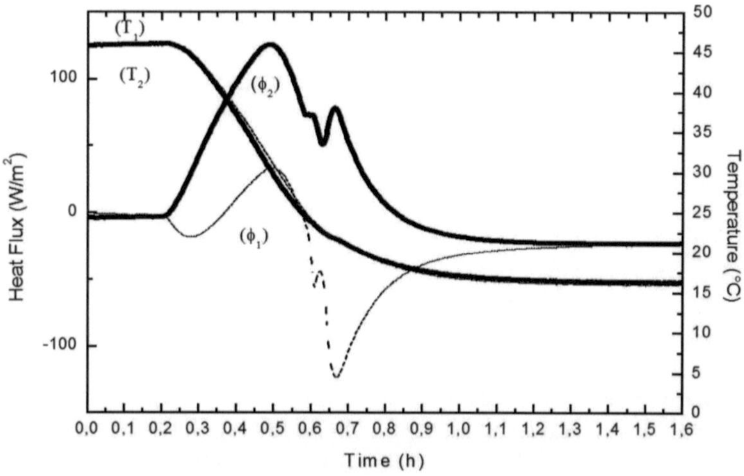

Figure 10.
Heat flux and temperatures evolutions (50–15°C) [6].

contact with the heat exchanger plate. This slight layer "insulates" the liquid PCM from the cooling source and reduce the effect of the convection mode. Then, the solidification continued slowly because of the low thermal conductivities of the solids SS-PCMs until the sample temperature reached 15°C at the end of the test. Furthermore, it can be noted that the period for solidification was much longer than the melting process.

4. Conclusion

Phase change materials (PCMs) with its capacity of storing thermal energy as latent heat is a viable approach of the utilization of solar heat, a green source of energy, and the optimization of energy consumption in buildings. An accurate material characterization of thermal energy storage materials including energy storage capacity and phase change temperature, during several charging and discharging cycles is significant to develop an efficient thermal storage device or application. Conventionally, thermal characterization of materials is performed using thermal analysis techniques especially Differential Scanning Calorimetry (DSC). However, all these methods involve very small samples that can be significantly influenced by local heterogeneities, the response dependence on the used heating rate, etc. For investigation of PCMs in real conditions, it is essential to design an experimental device to provide robust insights on the steady-state parameters, the dynamic thermal responses, the heat storage capacity of PCM at large scale, to meet the design requirements of thermal energy storage applications and to prevent potential failures. This chapter reports on successful use of a specific experimental method to measure the temperatures and the heat fluxes in order to characterize the phase change effects. The Transient Guarded Hot Plate Technique (TGHPT) provides reliable information about thermal conductivity, fusion enthalpy, specific heat and thermal stability for large phase change materials samples. Heat flux measurements make it possible to highlight very specific behaviors of these products and are thus a very interesting experimental source of data which comes to complete the traditional measurement methods like calorimetric device (differential scanning calorimetry). The Transient

Guarded Hot Plate Technique (TGHPT) allows large samples analysis and entails less maintenance, equipments price and time evaluation than the DSC. Thus, it is the most suitable method for PCM characterization.

Nomenclature

T	Temperature (°C)
C_P	Specific heat capacity (J/g.°C)
λ	Thermal conductivity (kJ/kg)
T_{Onset}	melting onset temperature (°C)
T_{endset}	melting endset temperature (°C)
T_m	melting temperature (°C)
ΔH_m	latent heat of fusion or melting enthalpy (kJ/kg)
T_{range}	temperature range of phase transformation (°C)

Subscripts and abbreviation

TES	Thermal Energy Storage
LHTES	Latent Heat Thermal Energy Storage
PCMs	Phase Change Materials
SS-PCMs	Shape-Stabilized PCMs
TGHPT	Transient Guarded Hot Plate Technique
DSC	Differential Scanning Calorimetry
DTA	Differential Thermal Analysis
init	initial thermal steady state
end	final thermal steady state
sens	sensible
s	solid state
l	liquid state
m	melting

Author details

Abdelwaheb Trigui
Laboratoire des Matériaux Multifonctionnels et Applications (LaMMA),
Université de Sfax, Tunisie

*Address all correspondence to: abdotrg@yahoo.fr

References

[1] Solé A, Miró L, Barreneche C, Martorell I, Cabeza LF. Review of the T-history method to determine thermophysical properties of phase change materials (PCM). Renewable and Sustainable Energy Reviews. 2013;**26**: 425-436. DOI: 10.1016/j.rser.2013.05.066

[2] Stritih U. An experimental study of enhanced heat transfer in rectangular PCM thermal storage. International Journal of Heat and Mass Transfer. 2004; **47**(12–13):2841-2847. DOI: 10.1016/j.ijheatmasstransfer.2004.02.001

[3] Trigui A, Karkri M, Boudaya C, Candau Y, Ibos L, Fois M. Experimental investigation of a composite phase change material: Thermal-energy storage and release. Journal of Composite Materials. 2014;**48**(1):49-62. DOI: 10.1177/0021998312468185

[4] Azzouz K, Leducq D, Gobin D. Enhancing the performance of household refrigerators with latent heat storage: An experimental investigation. International Journal of Refrigeration. 2009;**32**(7):1634-1644. DOI: 10.1016/j.ijrefrig.2009.03.012

[5] Nkwetta DN, Haghighat F. Thermal energy storage with phase change material—A state-of-the art review. Sustainable Cities and Society. 2014;**10**: 87-100. DOI: 10.1016/j.scs.2013.05.007

[6] Trigui A, Karkri M, Krupa I. Thermal conductivity and latent heat thermal energy storage properties of LDPE/wax as a shape-stabilized composite phase change material. Energy Conversion and Management. 2014;**77**:586-596. DOI: 10.1016/j.enconman.2013.09.034

[7] Richardson MJ. Quantitative aspects of differential scanning calorimetry. Thermochimica Acta. 1997;**300**(1–2): 15-28. DOI: 10.1016/S0040-6031(97)00188-3

[8] Feng G et al. DSC test error of phase change material (PCM) and its influence on the simulation of the PCM floor. Renewable Energy. 2016;**87**:1148-1153. DOI: 10.1016/j.renene.2015.07.085

[9] Castellón C, Günther E, Mehling H, Hiebler S, Cabeza LF. Determination of the enthalpy of PCM as a function of temperature using a heat-flux DSC-A study of different measurement procedures and their accuracy. International Journal of Energy Research. 2008;**32**(13):1258-1265. DOI: 10.1002/er.1443

[10] Mehling H, Cabeza LF. Heat and Cold Storage with PCM. Berlin, Heidelberg: Springer Berlin Heidelberg; 2008

[11] Günther E, Hiebler S, Mehling H, Redlich R. Enthalpy of phase change materials as a function of temperature: Required accuracy and suitable measurement methods. International Journal of Thermophysics. 2009;**30**(4): 1257-1269. DOI: 10.1007/s10765-009-0641-z

[12] Sarı A, Karaipekli A. Preparation, thermal properties and thermal reliability of capric acid/expanded perlite composite for thermal energy storage. Materials Chemistry and Physics. 2008;**109**(2-3):459-464. DOI: 10.1016/j.matchemphys.2007

[13] Solé A, Neumann H, Niedermaier S, Cabeza LF, Palomo E. Thermal stability test of sugar alcohols as phase change materials for medium temperature energy storage application. Energy Procedia. 2014;**48**:436-439. DOI: 10.1016/j.egypro.2014.02.05

[14] Neumann H, Niedermaier S, Gschwander S, Schossig P. Cycling stability of d-mannitol when used as phase change material for thermal storage applications. Thermochimica Acta. 2018;**660**:134-143. DOI: 10.1016/j.tca.2017.12.026

[15] Li T, Yuan Y, Zhang N. Thermal properties of phase change cement board with capric acid/expanded perlite form-stable phase change material. Advances in Mechanical Engineering. 2017;**9**(6): 168781401770170. DOI: 10.1177/1687814017701706

[16] Zhang S, Feng D, Shi L, Wang L, Jin Y, Tian L, et al. A review of phase change heat transfer in shape-stabilized phase change materials (ss-PCMs) based on porous supports for thermal energy storage. Renewable and Sustainable Energy Reviews. 2021;**135**:110127. DOI: 10.1016/j.rser.2020.110127

[17] Wu S, Yan T, Kuai Z, Pan W. Thermal conductivity enhancement on phase change materials for thermal energy storage: A review. Energy Storage Materials. 2019. DOI: 10.1016/j.ensm.2019.10.010

[18] Rathgeber C, Schmit H, Hennemann P, Hiebler S. Calibration of a T-history calorimeter to measure enthalpy curves of phase change materials in the temperature range from 40 to 200 °C. Measurement Science and Technology. 2014;**25**(3):035011. DOI: 10.1088/0957-0233/25/3/035011

[19] Bitter H, Lackner S. Fast and easy quantification of semi-crystalline microplastics in exemplary environmental matrices by differential scanning calorimetry (DSC). Chemical Engineering Journal. 2021;**423**:129941. DOI: 10.1016/j.cej.2021.129941

[20] Trigui A et al. Thermal and mechanical properties of maize fibres–high density polyethylene biocomposites. Journal of Composite Materials. 2013;**47**(11):1387-1397. DOI: 10.1177/0021998312447648

[21] Trigui A, Karkri M, Boudaya C, Candau Y, Ibos L. Development and characterization of composite phase change material: Thermal conductivity and latent heat thermal energy storage. Composites. Part B, Engineering. 2013; **49**:22-35. DOI: 10.1016/j.compositesb.2013.01.007

[22] Moulahi C, Trigui A, Karkri M, Boudaya C. Thermal performance of latent heat storage: Phase change material melting in horizontal tube applied to lightweight building envelopes. Composite Structures. 2016; **149**:69-78. DOI: 10.1016/j.compstruct.2016.04.011

[23] Chriaa I, Trigui A, Karkri M, Jedidi I, Abdelmouleh M, Boudaya C. Thermal properties of shape-stabilized phase change materials based on low density polyethylene, hexadecane and SEBS for thermal energy storage. Applied Thermal Engineering. 2020;**171**:115072. DOI: 10.1016/j.applthermaleng.2020.115072

[24] Moulahi C, Trigui A, Boudaya C, Karkri M. Smart macroencapsulated resin/wax composite for energy conservation in the built environment. Journal of Thermoplastic Composite Materials. 2017;**30**(7):887-914. DOI: 10.1177/0892705715614065

[25] Chriaa I, Karkri M, Trigui A, Jedidi I, Abdelmouleh M, Boudaya C. The performances of expanded graphite on the phase change materials composites for thermal energy storage. Polymer (Guildf). 2021;**212**:123128. DOI: 10.1016/j.polymer.2020.123128

Section 3

Application of Phase Change Material in Battery Thermal Management

Chapter 4

Review of the Use of the Carbon-Based Phase Change Material Composites in Battery Thermal Management for Electric Vehicles

Yang Yang

Abstract

With increasing attention to the environment issues, the replacement of traditional energy vehicles with new energy vehicles has gained support from more countries. Lithium battery is an energy storage component of an electric vehicle and hybrid vehicle. Due to the nature that lithium batteries are very sensitive to operating temperature, a battery thermal management system is required to improve its efficiency and life. This chapter discusses the technology of phase change materials in battery thermal management systems and reviews the performance characteristics and thermal properties of various types of lithium-ion batteries. Furthermore, it summarizes the breakthroughs and bottlenecks of carbon-based phase change material composites used in the lithium battery heat dissipation systems.

Keywords: Li-on batteries, electric vehicles, phase change material, thermal management, nanomaterials

1. Introduction

With the rapid development of the global economy, there is an increasing environmental problem caused by traditional fossil energy. Not only do the carbon dioxide and other harmful gasses produced by burning fossil fuels lead to the damage of the environment and the rise of global temperature, but they result in long-term risks to human health [1]. Thus, after the 2015 United Nations Climate Change Conference, most countries have reached a consensus to replace traditional energy with new one, reducing carbon emission. According to bp Statistical Review of World Energy 2021, the consumption of primary energy decreased by 4.5% worldly and the carbon emissions reduced by 6.3%, which was the largest drop since 1945 [2]. Due to the fact that transportation is one of the major causes of high carbon emissions, many countries have made a plan for a total ban on the sale of traditional fuel vehicles [3]. For example, the UK attempts to achieve zero carbon emissions of all new vehicles by 1940; France will implement a policy regarding a ban of traditional fuel and hybrid vehicle sales; India plans to inhibit the sales of petroleum fuel vehicles in 2030. Besides these countries, Spain, Norway, and the Netherlands have already set up a plan for banning the sale of traditional vehicles. Because the sales of traditional fuel vehicles are limited, the recent focus has shifted to new energy vehicles. Based on the Electric Vehicles Outlook 2021 published by BloombergNEF, passenger electric vehicle (EV) sales reached 3.1 million in 2020 and the sales of EV will continue to rise increasingly in the future, it is likely to reach 4 million in 2025, for instance [4].

At present, the development of new energy vehicles mainly focuses on electric vehicles. Based on the power source, electric vehicles are divided into three categories, battery electric vehicles (BEVs), hybrid electric vehicles (HEVs), and fuel cell vehicles (FCEVs) [5]. BEV technology uses a battery to drive an electric motor to generate power. Because there is no intervention of the internal combustion engine, it creates an environment where harmful gasses have zero emission during the working process. The increasingly popular kinetic energy recovery system (KERS) can use the brakes to charge the battery during driving, thereby expanding the drive range. HEVs rely on the work of both internal combustion engines and electric motors. HEVs have smaller engines than conventional cars, so it provides better fuel economy, effectively reducing the emissions of harmful gasses. FCEVs have not started in most countries due to some problems such as their high cost and immature technology. In the U.S., for example, the total sales of hydrogen FCEVs declined 12% to 2089 in 2019 [6].

Previous research and investigation showed that the increase in EVs, HEVs, and necessary power auxiliary facilities established on the basis of related policies can effectively reduce the emissions of harmful gasses and greenhouse gasses [7, 8]. However, the battery's limited temperature range and endurance has hindered the development of EVs and HEVs, which can only be realized by effectively improving the performance of battery pack and a large part of the improvement of the battery performance depends on thermal management.

2. Lithium batteries

Electric energy storage system (storage battery) is the energy storage component of EVs and HEVs, which determines the driving range of a vehicle. While using, storage batteries experience thousands of times of charging and discharging. Different types of batteries have various performances in energy density, battery life, and heat release.

EVs and HEVs have many types of batteries to choose from, such as nickel cadmium batteries, lead-acid batteries, and lithium batteries. Among them, lithium battery has higher energy density and stronger power than the others, which is why it has become the best choice of EVs and HEVs energy storage spare parts [9].

In respect of the low cost, cruising range, and safety requirements of EVs, lithium batteries are constantly innovating and developing. Current lithium battery cathode materials mainly contain lithium cobalt oxide (LiCoO2), lithium manganese oxide (LiMn2O4), lithium nickel cobalt manganese oxide (LiNiMnCoO2 or NMC), lithium iron phosphate (LiFePO4), etc. LiCoO2 is gradually being replaced by LiMn2O4 because of its short life, low thermal stability, and limited load. LiMn2O4 battery has a three-dimensional spinel structure, so it has the advantages of high stability and high safety. However, the disadvantage is that the charge and discharge cycle life is limited. Therefore, LiCoO2 and LiMn2O4 batteries are no longer used in the mainstream

Battery types	Specific energy (electric capacity) (Wh/Kg)	Charge rate (C)	Discharge rate (C)	Cycle life	Thermal runaway (°C)
LiCoO2	150–200	0.7–1	1	500–1000	150
LiMn2O4	100–150	0.7–1	1	300–700	250
LiNiMnCoO2	150–220	0.7–1	1	1000–2000	210
LiFePO4	90–120	1	1	1000–2000	270
LiNiCoAlO2	200–260	0.7	1	500	150
Li4Ti5O12	50–80	1–5	10	3000–7000	—

Table 1.
Properties of lithium batteries with different cathode materials [10, 11].

electric vehicle field. NMC batteries are composed of three active metals, nickel, cobalt, and manganese, mixed in different proportions to achieve large capacity, high load, and long life. It also has good overall performance and the lowest self-heating rate in the current battery technology, making it suitable as an energy storage unit for EVs. LiFePO4 battery has been used as a substitute for lead-acid battery for a long time and has become the energy storage unit of EVs. LiFePO4 has good electrochemical performance and low resistance. Its advantages include good thermal stability and safety, and long cycle of charge discharge under high rated current. Lithium titanate battery replaces the graphite in the anode of a typical lithium-ion battery and forms a spinel structure. The cathode of the battery can be LiMn2O4 or NMC. Lithium titanate batteries have outstanding performance in terms of safety, low temperature and longevity, but the disadvantage is the high cost and low capacitance. The properties of lithium batteries with different cathode materials are illustrated in **Table 1** [10, 11]. The types of anode materials are mainly divided into two categories, carbon-based and non-carbon-based. Carbon-based materials include graphite, carbon fiber, carbon nanotubes, graphene, etc. and non-carbon-based materials primarily include nitrides, tin-based materials, silicon-based materials, etc. Silicon-based and carbon-based materials are now considered to be the most promising anode materials.

When the actual vehicle is running, the operating temperature range of the battery pack is much wider than the optimal temperature range. The charging and discharging of the battery generates a lot of heat [12]. The temperature being too high or too low will affect the efficiency and life of lithium battery charging and discharging cycles and cause safety issues [13]. The battery pack is composed of several lithium batteries in series and parallel. Due to the uneven heat dissipation inside, the temperature of a certain area will be higher than the others, resulting in differences in the work efficiency and lifespan of individual lithium batteries inside the battery pack. Therefore, the battery thermal management system is an important part of the whole battery power system. It uses air cooling, liquid cooling, phase change materials, and other technologies to control the internal temperature of the battery pack under the strategy of the electronic control unit, so as to improve the working efficiency of the battery pack and prolong its operation.

3. PCM composites

The preparation of carbon-based PCM composite is divided into four steps, shown as the diagram in **Figure 1** [14]. To start with, myristic acid (MA) is heated

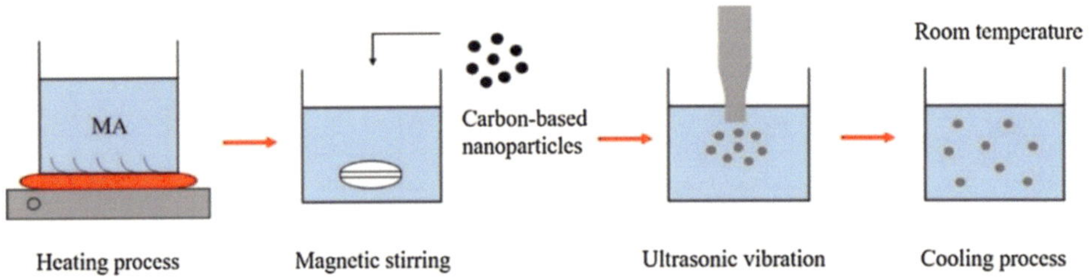

Figure 1.
Semantic diagram of sample preparation process [14].

above 70°C and melted completely for 10minutes [14]. Subsequently, carbon-based nanoparticles are added to MA while magnetic stirring for 30 minutes. Following this, the material needs ultrasonic vibration for 1 hour. After cooling down to room temperature, carbon-based PCM composite is prepared.

Reviewing the characteristics of the material mainly requires morphology analysis and chemical interaction examination. Specifically, the Scanning Electron Machine (SEM) is used to observe the microstructures of carbon-based PCM composite and Fourier transform infrared (FTIR) spectroscopy is employed to investigate the interaction between carbon additives and PCM. Thermophysical properties of the composite such as thermal conductivity is measured by thermal conductivity meter TC 300E. The relationship between enthalpy and temperature is explored by Differential Scanning Calorimetry (DSC) and the relationship between materials mass and temperature is examined via thermo gravimetric analysis.

Compared with the traditional air cooling system and liquid cooling system, the phase change material has the advantages such as simple structure, uniform temperature control, and no additional energy consumption. Its principle is to use the phase change process of the material at a specific temperature to absorb the heat energy generated by the battery in order to control the temperature. During the phase transition, though the temperature is constant or the temperature change range is small, it can absorb or release a large amount of latent heat [15]. Because the melting point of PCM is almost equal to the operating temperature, how to choose PCMs depends on the operating temperature of the lithium battery placed in the battery pack. The phase change temperature of PCMs that are generally used for battery thermal management is mainly concentrated at 30–50°C, and the temperature above 65°C or below 0°C has obvious adverse effects on its performance [16]. As the power of the battery pack increases, the volume and calorific value of the battery also rise, so the phase change material also needs to possess higher thermal conductivity and higher latent heat per unit mass to improve the optimal temperature control time. Moreover, PCMs should have other properties, such as the stable thermal cycle life, low density, safe, and non-toxic and inert chemical components, etc. PCMs can be roughly divided into two categories, organic materials and inorganic materials. **Table 2** shows the advantages and disadvantages of the two types of PCMs [17].

El Idi et al. focused their study on the topic concerning Li-ion cell thermal behavior and the optimization of a passive thermal management system utilizing a PCM-Metal Foam composite [18]. The material they used involved an aluminum foam/paraffin RT27 composite and Li-ion 18,650 cell, which was kept below 27°C. To test the efficiency of the passive thermal management system, they developed a two-dimensional numerical model, building on the enthalpy-porosity model and non-equilibrium equation. According to the results, they concluded that several factors

	Pros	Cons
Organic PCMs	Wide distribution of phase transition temperatures	Low latent heat of phase per unit volume
	No supercooling and phase separation	Low thermal conductivity in solid state
	Chemically and thermally stable	Volume changes during phase transition
	Good compatibility with traditional materials	
Inorganic PCMs	High latent heat of phase per unit volume	Severe supercooling problems and prone to phase separation
	High thermal conductivity	Corrosive and thermally unstable
	No volume change during phase transition	

Table 2.
The advantages and disadvantages of the two types of PCMs [17].

including current rate, duration charge and discharge cycle, and ambient temperature influence the temperature of the cell surface. Although in the solid–liquid phase change stage, the PCM absorbed the heat produced by the cell in latent form, its efficiency was restricted by PCM's low thermal conductivity. To increase the efficiency of the thermal management of a Li-ion cell, an aluminum foam can be added. They also pointed out that the temperature of the cell surface was not significantly affected by an excessive amount of PCM added.

Heyhat et al. compared different composites used in the battery thermal management system, including PCM and porous metal foam, PCM and nanoparticles, and PCM and fin [19]. The PCM used was n-eicosane and its properties are shown in **Table 3**. From the table, it could be said that the melting temperature of n-eicosane was 309.55 K. The initial temperature of the battery was 298.15 K was reasonably assumed. The metal foam, nanoparticles, and fins were copper. In 4.6 W and 9.2 W heat generation rates, adding nanoparticles to PCM insignificantly affected the performance of the system. Contrasting nanoparticles with fins somehow improved the performance with three fins performing better than 5 fins. Among these three composites, the combination of metal foam and PCM performed more efficiently, resulting in the decline of battery mean temperature by 4 to 6 K.

Karimi et al. studied the performance of thermal battery management systems after adding metal matrix and metal nanoparticles including Cu, Ag, and $Fe3O4$ to paraffin composite PCM [20]. The results showed that the addition of both metal matrix and metal nanoparticles improved thermal conductivity. Among these metal nanoparticles, Ag nanoparticles had the best performance. Particularly, it had the lowest body temperature, while metal matrix-PCM composite achieved the least temperature difference.

Jilte et al. used nano-enhanced PCM arranged in trans-radial and trans-axial multi-layer to enhance heat transfer [21]. The trans-radial configuration was a cell in two coaxial cylindrical containers, where PCM1 and PCM2 were filled in the containers adjacent to each cell respectively, whereas equal volume PCM1 and PCM2 were arranged in trans-axial configuration. To accommodate the same volume of PCM1 and PCM2, the trans-axial structure was preferred for battery modules. Two different arrangements were employed in this study, a two-layer 7 × 7 × 1 arrangement consisted of seven battery cells, seven nePCM1 containers, and one nePCM2 container, and a two-layer 7 × 1 × 1 arrangement comprised of seven cells, one nePCM1

	Density ($g\,cm^{-3}$)	Latent heat of fusion (kJ/kg)	Melting point (K)	Molecular weight (g/mol)	Specific heat capacity (J/(kg•K))	Thermal conductivity ($W\,m^{-1}\,K^{-1}$)	Thermal expansion coefficient (1/K)
Solid n-eicosane (298.15 K)	910	248	309.55	282.54	1926	0.423	—
Liquid n-eicosane (at 323.15 K)	769	—	—	282.54	2400	0.146	8.161 × 10–4

Table 3.
Properties of solid and liquid n-eicosane [19].

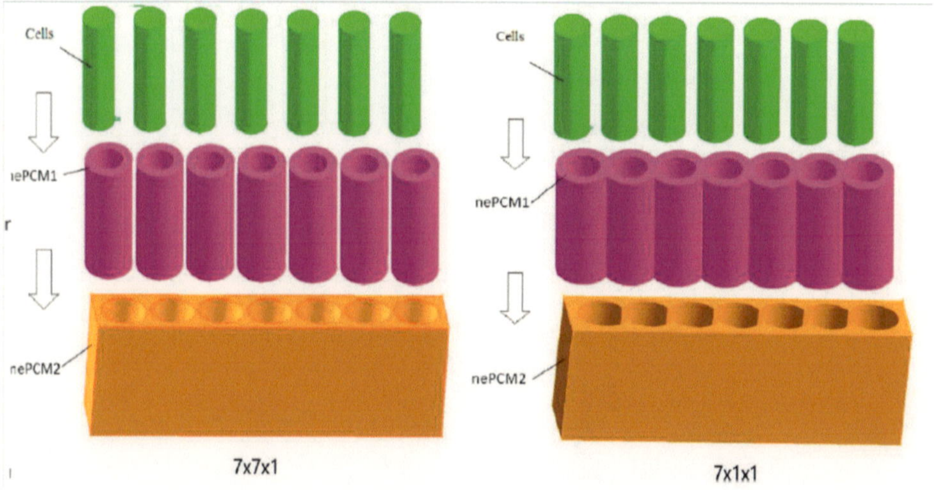

Figure 2.
7 × 7 × 1 and 7 × 1 × 1 arrangement [21].

container, and one nePCM2 container, as shown in **Figure 2**. In order to improve thermal conductivity, nanoparticles Al2O3 were doped with PCM1 Na2SO4•H2O and PCM2 eicosane. Their thermophysical properties are illustrated in **Table 4**. As shown in the table, under the same melting temperature, Na2SO4•H2O and eicosane had high thermal conductivity, which were preferred in the experiment. Based on the experiment conducted under the condition where the battery was at a 3C discharge rate and the ambient temperature was 30°C, they found that the 7 × 7 × 1 configuration was more efficient compared to 7 × 1 × 1 arrangement for improved battery cooling and the container next to the battery surfaces should be filled with nePCMs with a smaller value of melting temperature.

Lv et al. developed a novel material, a nanosilica (NS)-enhanced composite PCM (CPCM-NS), to overcome the obstacles such as PCM leakage and volume changes presented by many PCM cooling technologies [22]. This kind of composite is anti-leakage and anti-volume-change as NS has nanoscale pores that absorb liquid paraffin, which prevents PCM leakage and diminish volume change. CPCM-NS was prepared through melting paraffin in an oil bath at 85°C for an hour and adding different amounts of NS, 7 wt% EG, and 30 wt% low density polyethylene to the paraffin respectively. The entire preparation process is shown in the form of a diagram in **Figure 3**. It could be concluded from the results that CPCM-NS helped the battery achieve better cooling efficiency and durability. At the first, second, fourth, sixth, and eighth charge and discharge cycle, the maximum temperatures of CPCM-NS with 5.5 wt% of NS reached were 1.6, 2.4, 4.5, 5.3, and 5.9 lower than CPCM without NS. For the remaining cycles, the temperature difference was 6.22 ± 0.05°C.

Wei et al. chose cellulose nanocrystals (CNCs), graphene nanoplatelets (GNPs), and polyethylene glycol (PEG) as their PCM composite for GNPs offered thermally conductive path and CNCs restricted the leakage of PEG in the phase change process [23]. 8 wt% CNC was added to the composite, while four different types of GNP were added, 0.5 wt%, 1 wt%, 2 wt%, and 4 wt%. When GNPs were at 4 wt%, the enthalpy of the composite was 145.4 J/g and the thermal conductivity was 2.018 + −0.067 W/m K, proving that CNC enhanced the composite's enthalpy even when the thermal conductivity remained high. Additionally, this type of composite

Base PCM	Nanoparticles	Composite PCM	melting temperature (°C)	Latent heat (J/kg)	Specific heat (J/kgK)	Thermal conductivity (W/mK)	Density	Nanoparticles diameter (nm)
Disodium sulfate decahydrate(Na_2SO_4•H_2O)			36.4	254	1930	0.544	1485	
Eicosane			36.4	241	1900	0.27	778	
	Aluminum oxide (Al_2O_3)				765	36	3600	59

Table 4.
Thermophysical properties of PCM and nanoparticles [21].

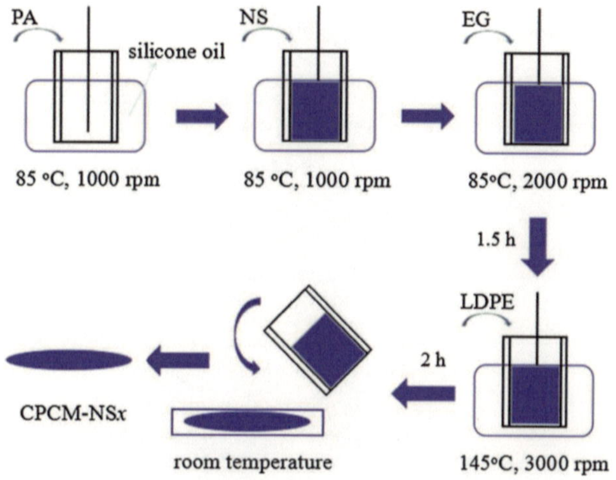

Figure 3.
Schematic diagram for the preparation of NS-enhanced CPCMs [22].

	Melting onset (°C)	Peak temperature (°C)	Latent heat (J/g)
K20 (20 wt% kaolin)	39.08	42.26	158.14
K15E5 (15 wt% kaolin, 5 wt% EG)	37.66	42.15	160.57
K12E8 (12 wt% kaolin, 8 wt% EG)	37.72	41.81	162.66
K10E10 (10 wt% kaolin, 10 wt% EG)	37.87	41.64	165.21
K8E12 (8 wt% kaolin, 12 wt% EG)	38.04	41.51	169.55

Table 5.
Properties of kaolin, EG, and paraffin composite [24].

displayed outstanding light to heat and electricity to heat energy conversion abilities based on the fact that the composite had high latent heat and retention rate.

Zhang et al. utilized kaolin, EG, and paraffin as PCM composite [24]. Among these three materials, kaolin and EG had different percentages. Their properties are shown in **Table 5**. The latent heat of these materials did not differentiate greatly from one to another considering the invariance of the mass fraction of paraffin. It was found that the bigger amount of EG is needed for a higher thermal conductivity of the composite and K20 and K15E were not desirable materials for they showed paraffin leakage. Despite the same thermal conductivity it had as K10E10, K8E12 was not selected due to the high cost and inadequate availability. Thus, K10E10 was used as the suitable material. The results indicated that the maximum temperature decreased by 13.4% at 2C discharge rate, 20.76% at 3C, and 27.74% at 4C under 26–28°C. The temperature difference was maintained at 4.04°C at 4C discharge rate, lower than the temperate difference when PCM was not used.

Lv et al. created serpentine CPCM (S-CPCM) plates to enhance secondary heat dissipation capability [25]. In doing so, the S-CPCM plates provide a larger surface area and more air flow channels. The S-CPCM plates utilized 5 wt% of expanded graphite and low-density polyethylene (LDPE) of differentiated wt%, ranging from 0 to 25wt%, which was added to the resultant CPCM. The properties including,

Base PCM	Nanoparticles	Composite PCM	Melting temperature (°C)	Latent heat (J/kg)
The PCM composed by hexadecane stearic acid and paraffin(11:1)			49.6	187.6
		S-CPCM+0% (low-density polyethylene)	49.6	172
		S-CPCM+3.5% (low-density polyethylene+3.5%Expanded graphite)	49.8	157.1
		S-CPCM+10.5% (low-density polyethylene+10.5%Expanded graphite)	50.2	142.2
		S-CPCM+17.5% (low-density polyethylene+17.5%Expanded graphite)	50.1	130.1
		S-CPCM+25% (low-density polyethylene+25%Expanded graphite)	50.3	119.4

Table 6.
Properties of S-CPCM of different composite [25].

the onset melting temperature, latent heat, and bending strength are indicated in **Table 6**. Based on the properties, S-CPCM-25% was eventually chosen for constructing the S-CPCM module for that S-CPCM-25% had suitable onset melting temperature, thermal conductivity, latent heat, and bending strength compared to S-CPCM of other types. The results from the experiment demonstrated that the S-CPCM structure effectively reduced the weight of the CPCM module by approximately 70% and provided larger heat exchange surface and more air convection channels, implying that the S-CPCM module performed better in heat dissipation than the traditional CPCM module. The positive aspect of the S-CPCM module provided a potential PCM cooling structure for lightweight battery modules.

Ling et al. applied a 60 wt% RT44HC/expanded graphite (EG) composite and a 60 wt% RT44HC/fumed silica composite to investigate the performance of these two materials at 5 and −10°C [26]. RT44HC is an organic PCM produced by Rubitherm Technologies GmbH (Rubitherm, n.d.). Its properties and the two composite materials' properties are included in **Table 7**, which showed similar phase change temperature and phase change enthalpy of the composites, but different thermal conductivity of them. The battery pack is composed of 20 cells and discharged at 0.5C, 1C, 1,5C, and 2C over 20 charge–discharge cycles. It was found that PCMs were effective in keeping the battery pack from cooling too fast. The RT44HC/fumed silica composite was superior in extending the cooling period to the RT44HC/EG composite. However, it was not suitable for a multi-cell battery pack for its low thermal conductivity contributed to a higher than 12°C temperature difference and a high voltage difference between battery cells. The RT44HC/EG composite was also successful in inhibiting battery overheat. They found the maximum temperature difference was 6°C. Furthermore, the RT44HC/EG composite improved temperature uniformity, further reducing the voltage differences.

Hussain et al. combined paraffin, chosen as their PCM, with graphene coated nickel (GcN) foam to study thermal management systems [27]. Four other materials were also compared including nickel foam, paraffin wax, GcN foam, and nickel foam saturated with paraffin. Among these materials, under 1.7A discharge current, GcN foam decreased the temperature rise of battery surface 17% compared to nickel foam. This novel material also improved the thermal conductivity. The results showed that

Base PCM	Composite PCMs	Phase change temperature (°C)	Specific heat capacity (J/g/°C)	Thermal conductivity (W/(m K))	Phase change enthalpy (J/g)
RT44HC		43.1	2.5	0.24	232.0
	60 wt% RT44HC/EG	42.8	2.4	9.57	134.3
	60 wt% RT44HC/fumed silica	41.5	0.8	0.18	133.4

Table 7.
Properties of RT44HC, 60 wt% RT44HC/EG composite, and 60 wt% RT44HC/fumed silica composite [26].

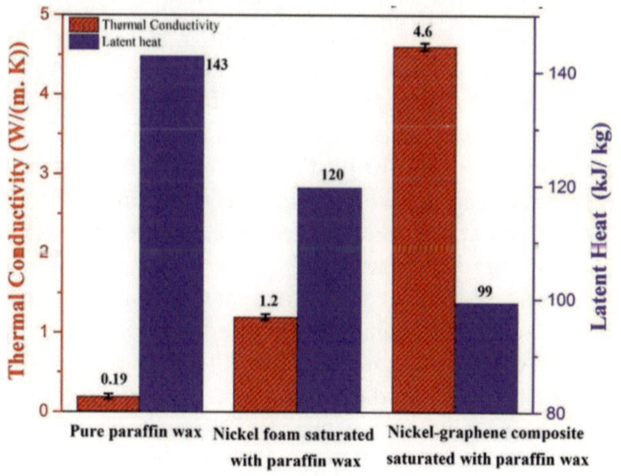

Figure 4.
Thermal conductivity (at 25°C) and latent heat paraffin composites [27].

the thermal conductivity of paraffin was improved by 23 times whereas that was improved by 6 times when using the nickel foam. **Figure 4** evidently showcases these results. Moreover, compared to pure paraffin, GcN foam's latent heat declined by 30% and its specific heat declined by 34%.

Wu et al. developed a battery thermal management system using thermally induced flexible composite PCM (FCPCM) [28]. FCPCM is composed of paraffin, olefin block copolymer (OBC), and EG. OBC and EG were mixed with paraffin correspondingly and their properties are illustrated in **Table 8**. Their properties suggested a good compatibility of these three materials as their enthalpy was close to the theoretical values. Since the storage modulus decreased from 1081 to 63.9 MPa when temperature rose from 25–60°C, FCPCM has good compatibility and flexibility. When the battery is discharged to 0%, the temperature of FCPCM reaches 43.4°C at 2.5C, 28.8°C lower compared to PCM were not used. This being said, the battery thermal management system with FCPCM performed well at thermal control, which was a result of low thermal contact resistance between the battery and FCPCM. Additionally, the results also suggested that FCPCM had lower temperature difference within the acceptable range and long-time function of latent heat.

	Phase change temperature (°C)	Thermal conductivity ($W\,m^{-1}\,K^{-1}$)	Phase change enthalpy ($J\,kg^{-1}$)
PA and OBC	39.82	0.45	200.6
PA, OBC, and EG	39.50	2.34	185.4

Table 8.
Properties of PA and OBC composite and PA, OBC, and EG composite [28].

Miers and Marconnet designed PCM heat sinks with three kinds of PCM composites for enhanced passive thermal management [29]. The PCMs were chosen based on their thermophysical properties such as melting temperature, thermal conductivity, and heat capacity. Eventually, PT42 supplied by PureTemp, PT 68 supplied by PureTemp, and S70 supplied by PlusICE were selected by PCM Products Ltd., n.d.; PureTemp LLC, n.d.). These materials' properties are shown in **Table 9**. In the table, an important property called figure of merit (FoMq) was included for it indicated the storage potential and the easiness of heat to be added or removed when selecting the PCM. It can be seen that S70 had the highest FoMq, suggesting it might be a more suitable material for the PCM. Based on the results, it was concluded that the isokite package design and S70 performed the best among other composites. In particular, at 7.5 W cm^{-2} heat flux, this design extended the time spent on reaching 95°C by 36.2% compared to S70 with a solid package. Furthermore, the weight of this design was 17.3% less than the solid aluminum package.

Zhang et al. developed an innovative flame-retarded composite comprised of paraffin, EG, ammonium polyphosphate (APP), red phosphorus (RP), and epoxy resin (ER) [30]. At a 3C discharge rate under 25°C, the composite reduced the peak temperature by 44.7% and 30.1% and controlled the maximum temperature difference within 1.36°C. Under high temperature of 45°C, the temperature uniformity was maintained within 5 °C. These results suggested that this novel composite PCM had outstanding thermophysical properties, improving the effectiveness of the thermal management system.

Huang et al. utilized Styrene butadiene styrene (SBS), paraffin, and aluminum nitride (AlN) as PCM composite to study the battery thermal management system [31]. Five types of composites with different ratios of paraffin, SBS, and AlN were compared. Based on the properties of different materials shown in **Table 10**, CPCM#3 was selected as it had better flexibility, smaller temperature difference and higher thermal conductivity, enhancing the temperature uniformity. Out of the three materials of the PCM composite, SBS has a unique structure to support the composite, AlN contributes to the stability and thermal conductivity of the composite, and paraffin has high latent heat. At 3C discharge rate, the maximum temperature reached was 42.4°C and the temperature difference was 9.2°C. Compared to other types of the composites, CPCM#3 helped the maximum temperature stay at 48.4°C at 3C discharge rate after nine cycles, and the temperate difference was 8.7°C, suggesting it had better heat dissipation performance and temperature uniformity.

Yan et al. focused specifically on the performance of paraffin and EG PCM composite [32]. Three types of paraffin wax of different phase change temperatures were used in the experiment and their properties are illustrated in **Table 11**. Their different properties gave rise to their different thermal performances. The results indicated that the PCM system had better cooling performance than the natural convection system especially when the cycling rate was high since the PCM composite absorbed a

	Melting temperature (°C)	Figure of merit (×103W^2•s/m4•K)	Latent heat of fusion (kJ/kg)	Density (kg/m3)	Specific heat (liquid: kJ/kg; solid: kJ/kg•K)	Thermal conductivity (W/m•K)	Sensible heat storage per unit mass (ΔEsens, 20 → 90°C, kJ/kg)
PT42 (organic)	42	27.80	218	Liquid: 850 Solid: 940	Liquid: 2.12 Solid: 1.98	Liquid: 0.15 Solid: 0.25	155.92
PT68 (organic)	68	28.71	220	Liquid: 870 Solid: 960	Liquid: 2.38 Solid: 2.04	Liquid: 0.15 Solid: 0.25	162.18
S70 (inorganic)	70	105.34	110	Liquid: 1680 Solid: –	Liquid: 2.10 Solid: –	Liquid: 0.57 Solid: –	157.50

Table 9.
Properties of PT42, PT68, and S70 [29].

	Thermal conductivity (W/m•K)	Phase transition temperature (Tp) (°C)	Phase change enthalpy (ΔH) (J/g)
CPCM#0 (50% paraffin, 50% SBS, and 0% AlN)	0.26	47.58	78.07
CPCM#1 (45% paraffin, 50% SBS, and 5% AlN)	0.38	48.32	76.83
CPCM#2 (40% paraffin, 50% SBS, and 10% AlN)	0.47	48.47	73.98
CPCM#3 (35% paraffin, 50% SBS, and 15% AlN)	0.51	46.82	57.06
CPCM#4 (30% paraffin, 50% SBS, and 20% AlN)	0.53	45.88	39.11

Table 10.
Properties of CPCM#0, CPCM#1, CPCM#2, CPCM#3, CPCM#4 [31].

	Phase change temperature (°C)	Thermal conductivity (W/m•K)	Latent heat (kJ/kg)
RT36	36	0.37	21.4
RT45	45	0.22	111.3
RT58	58	0.16	163.3

Table 11.
Properties of RT36, RT45, and RT58 [32].

substantial amount of heat throughout the phase change period. The researchers also found out that RT45 had better performance in the dynamic cycling, so the optimal phase change temperature of PCM composite was 45°C. The cooling performance of the PCM composite could be further improved by increasing the laying-aside time.

4. The combination of air cooling and liquid cooling

Malik el al. conducted a study to compare a thermal management system using phase change composite material with no cooling and liquid cooling system [33]. Graphene was combined with PCM to increase its thermal conductivity. The battery back consisted of LiFePO4 prismatic cells that were charged at 1C and discharged at 1C, 2C, 3C, and 4C. The results showed that the temperature gradient was significantly lower when the thermal management system contained phase change composite material compared to no cooling and liquid cooling. Specifically, the temperature declined by 20°C at a 4C discharge rate when 6 mm thick phase change composite plates were used. At a 1C discharge rate, the temperature decreased about 4°C, from 33.5°C to 29.1°C. These results implied that the thermal management system using phase change composite material could maintain the battery temperature in a certain range.

Yang et al. proposed a thermal model to evaluate the performance of the PCM/liquid combined cooling system with start-stop control [34]. To be specific, they calculated the battery heat generation by a semi-empirical equation and compared the cooling effects in different cooling schemes. Three factors including the PCM thickness, the channel width, and the coolant flow rate influenced the cooling effects. With respect to the thickness of PCM, due to the performance of huge latent heat and approximately isothermal phase change process, a reasonably thick PCM layer could improve the heat dissipation capacity of the cooling system and the temperature uniformity within the battery pack. The thickness of the PCM layer they chose was 1.5 mm for that not only could it enhance the cooling effects, but it could meet the premise of the energy density. The PCM/liquid coupled system they proposed was superior to the traditional liquid cooling system in terms of the start-stop frequency, the parasitic energy consumption, and the temperature control reliability. It reduced the start-stop frequency less than half, saving the power dissipation and reducing the pumping power consumption by at 60%. Additionally, the PCM layer had the early warning function, which allowed the PCM/liquid system to control the battery temperature better than the traditional liquid cooling scheme.

Akabarzadeh et al. proposed a new concept of an innovative liquid cooling plate for thermal management of Li-ion batteries by incorporating a phase change material inside [35]. The geometrical model of the plate is illustrated in **Figure** 5. The cooling plate the researchers utilized was hybrid as it offered active and passive cooling methods. In addition, the hybrid liquid cooling plate was 36% lighter than a traditional aluminum liquid cooling plate of equivalent volume and had the capability of slowing the temperature loss during the cold stop via heating solution. Through testing the prototype of the hybrid liquid cooling plate, it was found that compared to an aluminum plate, the hybrid liquid cooling plate lowered up to 30% of energy consumption of the pump for circulating the coolant. Not only did the hybrid liquid cooling plate increase the temperature uniformity, it could effectively reduce a fast temperature drop of the cooling plate, which further decreased the energy required for the active heating process after short-term parking. Therefore, the hybrid liquid cooling plate with lightweight structure can be a promising candidate for electric vehicle battery packs.

Chen et al. developed a hybrid-PCM-liquid cooling system to study a Li-ion battery module under fast charging [36]. The experiment was conducted under 25°C

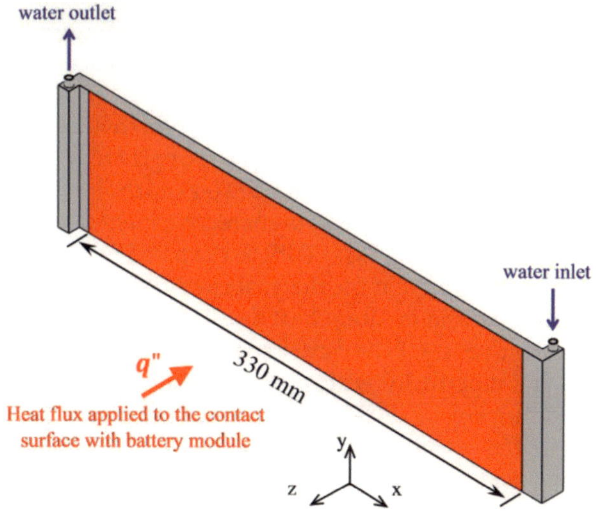

Figure 5.
Geometrical model of the liquid cooling plate [35].

and under 2C, 2.5C, and 3C fast charging rates, pure PCM, pure liquid, and hybrid-PCM-liquid cooling performance were investigated. Except for this, how the thickness of PCM and liquid coolant flow influenced the cooling effect was also analyzed. The results showed that temperature rise or distribution did not help improve each increment of PCM thickness and thick PCM affected the effectiveness of natural convection between the battery cell and the environment. Moreover, the attempt to achieve the optimization of thermal performance cannot be completed through the highest coolant flow rate and the thickest PCM. In contrast to pure PCM cooling systems and pure liquid cooling systems, hybrid-PCM-liquid cooling systems are capable of reaching temperature rise and distribution of acceptable range. Particularly, under 3C fast charging, 0.65 mm PCM and 54 mL/min coolant flow rate could enable the hybrid cooling system to achieve optimization and control the energy consumption within a suitable range.

Cao et al. combined PCM with delayed liquid cooling technology to achieve high temperature uniformity during high discharge rate [37]. The results showed that the delayed liquid cooling strategy decreased the temperature and temperature difference. It could even save energy consumption by shortening the period of liquid cooling. At 4C discharge rate with a cooling flow rate of 40 L/h, this design performed better than traditional strategy with continued liquid cooling.

Zhang et al. created a battery thermal management system with the combination of PCM and liquid cooling, taking advantage of the heat dissipation function this hybrid system presented [38]. They found that the PCM prevented thermal runaway diffusion under extreme conditions and the liquid cooling system effectively transmitted the heat PCM absorbed.

Wang et al. used different wt% OP28E nano-emulsions for a liquid cooling thermal management system. The results showed that the 10% OP28E nano-emulsion performed better than water and the increased mass fraction of OP28E resulted in decreased maximum temperature and maximum temperature difference [39]. Moreover, these two features were also related to the increase in coolant flow rate.

Fan et al. studied air cooling systems by adding metal fins to the PCM systems to enhance thermal control of the battery [40]. They found that the application of fins increased the working time of the battery by 98.4% compared to the battery-PCM system. Even under circumstances where the ambient temperature was high, the safe operation time of the system improved by 1.48 times for 20°C, 1.49 times for 30°C, and 1.81 times for 40°C.

Lv et al. employed EG, paraffin, and low density polyethylene (LDPE) coupled with low fins to improve the issues of traditional battery thermal management regarding PCM leakage and low surface heat transfer capability [41]. The LDPE-enhanced composite PCM (L-CPCM) was prepared via melting paraffin in an oil bath at 60°C for half an hour and adding EG, LDPE to the paraffin. Their properties are shown in **Table 12**. It was found that the thermal conductivity of L-CPCM is much higher than that of paraffin and its latent heat decreased greatly with the existence of EG and LDPE. Not only does L-CPCM have better bending strength, impact strength, and Shore hardness, but it performs well at heat dissipation. At 3.5C discharge rate, the maximum temperature of L-CPCM coupled with fins could remain below 50°C and the temperature difference was 5°C, while that L-CPCM without fins stayed up to 52.6°C and the temperature difference was 5.7°C. Thus, L-CPCM with fins could enhance the surface heat transfer capability.

Jiang et al. designed a battery pack with EG and paraffin composite as PCM coupled with a forced air cooling system, which consisted of aluminum tubes, baffles

	Density ($g\,cm^{-3}$)	Thermal conductivity ($W\,m^{-1}K^{-1}$)	Specific heat capacity ($J\,g^{-1}K^{-1}$)	Latent heat ($J\,g^{-1}$)	Phase change temperature (°C)
Paraffin	0.910	0.16	2.68	182	44.0–50.2
EG	0.230	8.0	0.71	—	—
LDPE	0.925	0.30	2.30	—	—
L-CPCM	0.856	1.38	2.48	87.4	44.5–50.2

Table 12.
Properties of paraffin, EG, LDPE, and L-CPCM [41].

and a shell [42]. The results indicated that the EG and paraffin composite reduced the temperature rising at 5C discharge rate and maintained the maximum temperature difference within 1 to 2 °C. The baffles applied improved the efficiency of heat transfer and the interaction of fluid by changing the air fluid flow direction.

Wu et al. studied the performance of PCM coupled with an air cooling system by using a heat pipe [43]. The forced art convection allowed the highest temperature to reach below 50°C at 5C discharge rate and the temperature difference to vary less. Moreover, the addition of a heat pipe significantly reduced the temperature compared to the usage of PCM only. Even after the first cycle, this module was able to maintain a stable stage with the same temperature profile as the first one.

Chen et al. investigated the influence of heat pipes and the thickness of PCM on the battery thermal management system [44]. They found that increased thickness of PCM and latent heat decreased the maximum temperature but increased the temperature difference. The heat dissipation could be improved via increasing the equivalent thermal conductivity of the heat pipe and reducing the temperature at start. The optimization of PCM's thickness could effectively improve the performance of the system without expanding the system volume.

Huang et al. investigated the performance of three battery thermal management system modules, including pure PCM, PCM coupled with air cooling, and PCM coupled with liquid cooling [45]. The results showed better performance of PCM coupled with a liquid cooling system than the other two modules. It was especially able to maintain the highest temperature at 50°C at 3C discharge rate and a lower temperature difference, which indicated it had outstanding temperature control and balance.

5. Conclusion

This paper discusses the classification, advantages, and disadvantages of lithium battery technology for electric vehicles, as well as the current status and application prospects of BTMS technology for composite phase change materials. The following conclusions have been drawn:

- At present, liFePO4 batteries are currently the most suitable batteries for electric vehicles. Lithium titanate batteries have excellent performance in terms of safety, low temperature, service life, etc. If there is a breakthrough in the improvement of its permittivity in the future, it will become the most promising lithium-ion battery.

- The use of composite PCM is currently the most potential thermal management technology for batteries. The fusion of carbon nanotubes and other composite materials can solve the problem of low thermal conductivity and the combined cooling technology can solve the issues of latent heat and thermal conductivity. The technology regarding the fusion of nanoparticles, such as silica, has also brought progress in properties including anti-leakage and reduced volume changes in phase transitions. However, there is still room for improvement in the compatibility of composite materials and stability in cyclic testing.

- To improve the thermal conductivity of PCMs, carbon materials could be added. However, carbon-based PCM composites tend to have relatively low latent heat compared to PCMs. Therefore, factors such as the pore size and mass fraction should be taken into consideration when balancing the thermal conductivity and latent heat of carbon-based PCM composites.

- There are various ways to enhance the performance of the battery thermal management systems, for example, the use of PCM and heat pipe, and air or liquid cooling techniques. PCM, heat pipe, and liquid cooling techniques are more efficient than air-forced systems. On the other hand, liquid cooling is difficult to maintain and has a short life of only 3 to 5 years. In terms of the cost, the application of a heat pipe costs more than that of PCMs. That being said, PCM is a better choice to promote the performance of the battery management systems.

Abbreviations

AIN	Aluminum nitride
APP	Ammonium polyphosphate
BEVs	Battery electric vehicles
BTMS	Battery thermal management system
CNCs	Cellulose nanocrystals
EG	Expanded graphite
EVs	Electric vehicles
ER	Epoxy resin
FCEVs	Fuel cell vehicles
FCPCM	Flexible composite PCM
GNPs	Graphene nanoplatelets
HEVs	Hybrid electric vehicles
KERS	Kinetic energy recovery system
LDPE	Low density polyethylene
NS	Nanosilica
OBC	Olefin block copolymer
PCM	Phase change material
PEG	Polyethylene glycol
RP	Red phosphorus
SBS	Styrene butadiene styrene

Author details

Yang Yang
University Putra Malaysia, Beijing, China

*Address all correspondence to: 306627023@qq.com

References

[1] Ahmadi P. Environmental impacts and behavioral drivers of deep decarbonization for transportation through electric vehicles. Journal of Cleaner Production. 2019;**225**:1209-1219. DOI: 10.1016/j.jclepro.2019.03.334

[2] Bp p.l.c. Bp Statistical Review of World Energy 2021. 2021. Available from: https://www.bp.com/content/dam/bp/business-sites/en/global/corporate/pdfs/energy-economics/statistical-review/bp-stats-review-2021-full-report.pdf [Accessed: June 2, 2022]

[3] Gordon D. The Role of Transportation in Driving Climate Disruption. 2010. Available from: https://carnegieendowment.org/files/transport_climate_disruption.pdf [Accessed: June 2, 2022]

[4] BloomberNEF. Electric Vehicle Outlook 2021: Executive Summary. 2021. Available from: https://bnef.turtl.co/story/evo-2021/page/1 [Accessed: June 2, 2022]

[5] Wikipedia. History of the Automobile. 2022. Available from: https://fr.wikipedia.org/wiki/Histoire_de_l%27automobile [Accessed: June 2, 2022]

[6] Kane M. In 2019 Sales of Hydrogen Fuel Cell Cars in the U.S. Decreased. 2020. Available from: https://insideevs.com/news/392360/2019-sales-hydrogen-fuel-cell-cars-us/ [Accessed: June 2, 2022]

[7] Conte FV. Battery and battery management for hybrid electric vehicles: A review. Elektrotechnik & Informationstechnik. 2006;**123**:424-431. DOI: 10.1007/s00502-006-0383-6

[8] Kampman B, Leguijt C, Bennink D, Wielders L, Rijkee X, Buck A, Braat W. Green Power for Electric Cars: Development of Policy Recommendations to Harvest the Potential of Electric Vehicles. 2010. Available from: https://cedelft.eu/wp-content/uploads/sites/2/2021/04/4037_defreportBKa_1265711066.pdf [Accessed: June 2, 2022]

[9] Morton I. Lithium-ion Batteries Take Lead in EV Tests. 2001. Available from: http://proxy.library.nyu.edu/login?url=http://search.ebscohost.com/login.aspx?direct=true&db=bth&AN=5668408&site=ehost-live [Accessed: June 2, 2022]

[10] Xie J, Chen Y, Liu Y, Liu D, Song S, Zhang X, et al. Current situation and development trend of cathode material of electric vehicle power lithium battery. Advanced Materials Industry. 2015;**11**:54-59. DOI: 10.3969/j.issn.1008-892X.2015.11.011

[11] Li F, Zhang X, Wu Y, Li Y. Current situation and development trend of power lithium battery industry. Energy Conservation & Environmental Protection in Transportation. 2016;**3**:14-16. DOI: 10.3969/j.issn.1673-6478.2016.03.004

[12] Pesaran AA. Battery thermal management in EVs and HEVs: Issues and solutions. In: Advanced Automotive Battery Conference; 6-8 February 2001. Las Vegas: Battery Man; 2001. pp. 34-49

[13] Li D, Yang K, Chen S, Wu F. Thermal behavior of overcharged nickel/metal hydride batteries. Journal of Power Sources. 2008;**184**:622-626. DOI: 10.1016/j.jpowsour.2008.03.059

[14] He M, Yang L, Lin W, Chen J, Mao X, Ma Z. Preparation, thermal

characterization and examination of phase change materials (PCMs) enhanced by carbon-based nanoparticles for solar thermal energy storage. Journal of Energy Storage. 2019;**25**:100874. DOI: 10.1016/j.est.2019.100874

[15] Sharma A, Tyagi VV, Chen CR, Buddhi D. Review on thermal energy storage with phase change materials and applications. Renewable and Sustainable Energy Review. 2009;**13**:318-345. DOI: 10.1016/j.rser.2007.10.005

[16] Al-Hallaj S, Selman JR. Thermal modeling of secondary lithium batteries for electric vehicle/hybrid electric vehicle applications. Journal of Power Sources. 2002;**110**:341-348. DOI: 10.1016/S0378-7753(02)00196-9

[17] Khudhair AM, Farid M. A review on energy conservation in building applications with thermal storage by latent heat using phase change materials. In: Farid M, Auckaili A, Gholamibozanjani G, editors. Thermal Energy Storage with Phase Change Materials. 1st ed. Boca Raton: CRC Press; 2021. pp. 162-175. DOI: 10.1201/9780367567699.ch3

[18] El Idi MM, Karkri M, Tankari MA. A passive thermal management system of Li-ion batteries using PCM composites: Experimental and numerical investigations. International Journal of Heat and Mass Transfer. 2021;**169**:120894. DOI: 10.1016/j.ijheatmasstransfer.2020.120894

[19] Heyhat MM, Mousavi S, Siavashi M. Battery thermal management with thermal energy storage composites of PCM, metal foam, fin and nanoparticle. Journal of Energy Storage. 2020;**28**:101235. DOI: 10.1016/j.est.2020.101235

[20] Karimi G, Azizi M, Babapoor A. Experimental study of a cylindrical lithium ion battery thermal management using phase change material composites. Journal of Energy Storage. 2016;**8**:168-174. DOI: 10.1016/j.est.2016.08.005

[21] Jilte R, Afzal A, Panchal S. A novel battery thermal management system using nano-enhanced phase change materials. Energy. 2021;**219**:119564. DOI: 10.1016/j.energy.2020.119564

[22] Lv Y, Situ W, Yang X, Zhang G, Wang Z. A novel nanosilica-enhanced phase change material with anti-leakage and anti-volume-changes properties for battery thermal management. Energy Conversion and Management. 2018;**163**:250-259. DOI: 10.1016/j.enconman.2018.02.061

[23] Wei X, Jin XZ, Zhang N, Qi XD, Yang JH, Zhou ZW, et al. Constructing cellulose nanocrystal/graphene nanoplatelet networks in phase change materials toward intelligent thermal management. Carbohydrate Polymers. 2021;**253**:117290. DOI: 10.1016/j.carbpol.2020.117290

[24] Zhang X, Liu C, Rao Z. Experimental investigation on thermal management performance of electric vehicle power battery using composite phase change material. Journal of Cleaner Production. 2018;**201**:916-924. DOI: 10.1016/j.jclepro.2018.08.076

[25] Lv Y, Liu G, Zhang G, Yang X. A novel thermal management structure using serpentine phase change material coupled with forced air convection for cylindrical battery modules. Journal of Power Sources. 2020;**468**:228398. DOI: 10.1016/j.jpowsour.2020.228398

[26] Ling Z, Wen X, Zhang Z, Fang X, Gao X. Thermal management performance of phase change materials with different thermal conductivities for Li-ion battery packs operated at low

temperatures. Energy. 2018;**144**:977-983. DOI: 10.1016/j.energy.2017.12.098

[27] Hussain A, Abidi IH, Tso CY, Chan KC, Luo Z, Chao CY. Thermal management of lithium ion batteries using graphene coated nickel foam saturated with phase change materials. International Journal of Thermal Sciences. 2018;**124**:23-35. DOI: 10.1016/j.ijthermalsci.2017.09.019

[28] Wu W, Liu J, Liu M, Rao Z, Deng H, Wang Q, et al. An innovative battery thermal management with thermally induced flexible phase change material. Energy Conversion and Management. 2020;**221**:113145. DOI: 10.1016/j.enconman.2020.113145

[29] Miers CS, Marconnet A. Experimental investigation of composite phase change material heat sinks for enhanced passive thermal management. Journal of Heat Transfer. 2021;**143**:013001. DOI: 10.1115/1.4048620

[30] Zhang J, Li X, Zhang G, Wu H, Rao Z, Guo J, et al. Experimental investigation of the flame retardant and form-stable composite phase change materials for a power battery thermal management system. Journal of Power Sources. 2020;**480**:229116. DOI: 10.1016/j.jpowsour.2020.229116

[31] Huang Q, Deng J, Li X, Zhang G, Xu F. Experimental investigation on thermally induced aluminum nitride based flexible composite phase change material for battery thermal management. Journal of Energy Storage. 2020;**32**:101755. DOI: 10.1016/j.est.2020.101755

[32] Yan J, Li K, Chen H, Wang Q, Sun J. Experimental study on the application of phase change material in the dynamic cycling of battery pack system. Energy Conversion and Management. 2016;**128**:12-19. DOI: 10.1016/j.enconman.2016.09.058

[33] Malik M, Dincer I, Rosen M, Fowler M. Experimental investigation of a new passive thermal management system for a Li-ion battery pack using phase change composite material. Electrochimica Acta. 2017;**257**:345-355. DOI: /10.1016/j.electacta.2017.10.051

[34] Yang N, Wang M, Wang J, Wang L, Fu Y. A model-based assessment of controllable phase change materials/liquid coupled cooling system for the power Lithium-ion battery pack. Energy Technology. 2021;**9**:2000924. DOI: 10.1002/ente.202000924

[35] Akbarzadeh M, Jaguemont J, Kalogiannis T, Karimi D, He J, Jin L, et al. A novel liquid cooling plate concept for thermal management of lithium-ion batteries in electric vehicles. Energy Conversion and Management. 2021;**231**:113862. DOI: 10.1016/j.enconman.2021.113862

[36] Chen S, Garg A, Gao L, Wei X. An experimental investigation for a hybrid phase change material-liquid cooling strategy to achieve high-temperature uniformity of Li-ion battery module under fast charging. International Journal of Energy Research. 2021;**45**:6198-6212. DOI: doi.org/10.1002/er.6241

[37] Cao J, Ling Z, Fang X, Zhang Z. Delayed liquid cooling strategy with phase change material to achieve high temperature uniformity of Li-ion battery under high-rate discharge. Journal of Power Sources. 2020;**450**:227673. DOI: 10.1016/j.jpowsour.2019.227673

[38] Zhang W, Liang Z, Yin X, Ling G. Avoiding thermal runaway propagation of lithium-ion battery modules by using hybrid phase change material and liquid

cooling. Applied Thermal Engineering. 2021;**184**:116380. DOI: 10.1016/j.applthermaleng.2020.116380

[39] Wang F, Cao J, Ling Z, Zhang Z, Fang X. Experimental and simulative investigations on a phase change material nano-emulsion-based liquid cooling thermal management system for a lithium-ion battery pack. Energy. 2020;**207**:118215. DOI: 10.1016/j.energy.2020.118215

[40] Fan R, Zheng N, Sun Z. Evaluation of fin intensified phase change material systems for thermal management of Li-ion battery modules. International Journal of Heat and Mass Transfer. 2021;**166**:120753. DOI: 10.1016/j.ijheatmasstransfer.2020.120753

[41] Lv Y, Yang X, Li X, Zhang G, Wang Z, Yang C. Experimental study on a novel battery thermal management technology based on low density polyethylene-enhanced composite phase change materials coupled with low fins. Applied Energy. 2016;**178**:376-382. DOI: 10.1016/j.apenergy.2016.06.058

[42] Jiang G, Huang J, Liu M, Cao M. Experiment and simulation of thermal management for a tube-shell Li-ion battery pack with composite phase change material. Applied Thermal Engineering. 2017;**120**:1-9. DOI: 10.1016/j.applthermaleng.2017.03.107

[43] Wu W, Yang X, Zhang G, Chen K, Wang S. Experimental investigation on the thermal performance of heat pipe-assisted phase change material based battery thermal management system. Energy Conversion and Management. 2017;**138**:486-492. DOI: 10.1016/j.enconman.2017.02.022

[44] Chen K, Hou J, Song M, Wang S, Wu W, Zhang Y. Design of battery thermal management system based on phase change material and heat pipe. Applied Thermal Engineering. 2021;**188**:116665. DOI: 10.1016/j.applthermaleng.2021.116665

[45] Huang Q, Li X, Zhang G, Zhang J, He F, Li Y. Experimental investigation of the thermal performance of heat pipe assisted phase change material for battery thermal management system. Applied Thermal Engineering. 2018;**141**:1092-1100. DOI: j.applthermaleng.2018.06.048

Chapter 5

Phase Transformation Processes in the Active Material of Lead-acid Batteries

Alberto F. Romero and Pilar Ocón

Abstract

The good performance of a lead-acid battery (LAB) is defined by the good practice in the production. During this entire process, PbO and other additives will be mixed at set conditions in the massing procedure. Consequently, an active material mainly composed of unreacted PbO, lead sulfate crystals, and amorphous species will be obtained. Later, the same mass will be pasted on the grids and the curing step will be performed. In this way, the previous pasted mass will be modified and a new hard porous structure will be formed in the active material. Furthermore, this structure will be bounded to the grid through a corrosion layer. Thus, the formed plate will be conducted to the following soaking and formation procedures. In these manufacturing steps, thanks to the major role of H_2SO_4, the active non-conductive material will be transformed into an electrically conductive element. Therefore, the prior compounds (PbO and lead sulfate crystals) will be converted to new phases: Pb or oxidized to PbO_2 on the negative and positive plate, respectively. Because of the importance of the previous phase transformations, new advanced designs are focused on the internal structure of the active material to improve the LAB performance.

Keywords: lead-acid battery, curing process, formation process, Pb nanoparticles

1. Introduction

Lead-acid batteries (LABs) are supported by a large and well-organized network of suppliers and manufacturers. Additionally, in terms of market, this type of device is recognized as the leader for automotive batteries and the second most important for industrial batteries. Nowadays, the systems known as advanced LABs are commonly used for commercial and domestic purposes. On the other hand, distinct technologies are found in LABs, such as the flooded or the valve-regulated sealed (VRLA) ones. Furthermore, two types of grating are manufactured either the flat pasted one or the tubular one. Unlike a standard battery, the negative plate of an advanced battery is modified in several ways. Thus, the plate can be composed of the active material and a supercapacitor (active carbon covering) or directly composed of a single supercapacitor. In addition to the supercapacitor, the carbon compounds are also used as negative plate additives in small contents. Nevertheless, the composition of the positive plate for either standard or advanced LABs is practically the same. Another interesting point about the LABs is their remarkable sustainability. Because 60% of world production is formed by batteries build through recycled lead [1].

The battery performance is very influenced by the several phase changes in the active material. These changes are mainly caused and controlled by the manufacturing steps. Therefore, the LAB manufacturing process is so important.

In the first production step, the PbO powder is used as a precursor compound and it is massed with other additives. As a note, the PbO specie can be found as a mixture of two polymorphic structures (α PbO and β PbO). Therefore, when the massing procedure is finished, new compounds are formed (lead sulfate crystals and amorphous species). As soon as the grids are pasted, the curing step is carried out. As a result, the hard porous structure of the mass is built. Once the plates are cured, they are formed. Nevertheless, there is a previous procedure where the cured plates are soaked in H_2SO_4 for a long period of time. Thus, the internal pores of the active material are filled with acid and several transformations into the active material are found. Finally, in the last step, the formation is developed to obtain electrically conductive plates. Hence, the negative plate is mainly composed of Pb and the positive plate of a mixture of α and β PbO_2, being the β phase more stable in acidic media.

After this entire process, the LAB can be used for a specific application. Therefore, during the battery lifetime, different reactions are performed by charge and discharge processes. The main ones are defined by the following systems: PbO_2/ $PbSO_4$ (positive plate) and Pb/$PbSO_4$ (negative plate). Nevertheless, these systems are followed by another secondary process. Some of them are the sulfation, the water consumption, and the self-discharge. Furthermore, in some cases, a battery failure can be originated from these secondary reactions [2].

Lastly, as mentioned previously, a good battery operation is influenced by the internal structure of the LAB plates. For this reason, in the last few years, different investigations have been developed to enhance the battery performance through studies focused on modifications of the active material [3–6].

2. Lead oxide as the precursor compound

The active material spent to build the LAB plates is essentially composed of PbO. This lead compound is obtained by thermal oxidation of metallic lead in previous manufacturing steps. The resulting product is not entirely PbO since a portion of metallic lead cannot be oxidized. In this way, the percentage of PbO content is usually around 80–20% for free Pb. In addition, PbO is composed of two polymorphic structures in nature: the red tetragonal PbO or the α PbO and the yellow orthorhombic PbO or the β PbO. Moreover, PbO is basically made up of α PbO and a little β PbO content (around 5%). Additionally, the massing procedure is affected by the prior polymorphic compounds. Since the composition of the active material will be influenced by them. In this way, the formation of tribasic lead sulfate ($3PbO{\cdot}PbSO_4{\cdot}H_2O$/3BS) is mostly caused by the α PbO. However, the collecting of tetrabasic lead sulfate ($4PbO{\cdot}PbSO_4$/4BS) is obtained by the β PbO. Nevertheless, the best 4BS production conditions were experimentally demonstrated as 80% α PbO + 20% β PbO [2].

Comparing the relevance of the 3BS and 4BS compounds, the 3BS one is more used in the LAB industry. Because its nature characteristics (particle size, structure, and agglomerates) are ideal for the production of both negative and positive plates. In addition, because of previous treatments, the 4BS preparation is more complicated and expensive than the 3BS one. For example, the 4BS formation is commonly produced by the 3BS curing above 80°C for 48–72 h [2].

3. Massing procedure

After the production of positive and negative active masses, the paste can be composed of unreacted PbO, lead sulfate crystals, and amorphous species, such as hydrated lead oxides, lead carbonates, and free lead. In the case of lead sulfate crystals, several species can be found: $PbSO_4$, monobasic lead sulfate ($PbO{\cdot}PbSO_4$/1BS), 3BS, and 4BS [2, 7].

During the massing, the paste composition is modified by several experimental parameters, such as the pH, the temperature, and H_2SO_4/PbO ratio.

On the one hand, early in the procedure (pH from 1.0 to 8.0), the H_2SO_4 is consumed, and the formation of $PbSO_4$ and 1BS is noticeable Eqs. (1, 2). However, a small amount of 3BS is also formed Eq. (3). Thus, with the progress of the procedure, the pH is increased (pH > 9.5). Under these new conditions, the formation of 3BS is carried out as the main reaction, while the solubility of PbO is greatly decreased. As a consequence, the formation of 3BS crystals is only found on the surface of the PbO particles and thus, part of the PbO is isolated. Consequently, a percentage of PbO is not reacted, although the H_2 SO_4 content was stoichiometrically correct. Hence, according to their stability, distinct lead sulfate crystals are found as the predominant compound at three clear pH regions: 3BS at pH > 9.5, 1BS at 8.5 < pH < 9.5 and $PbSO_4$ at pH > 8.5 [2, 8].

$$PbO + H_2SO_4 \leftrightarrow PbSO_4 + H_2O \tag{1}$$

$$2PbO + H_2SO_4 \leftrightarrow PbO \cdot PbSO_4 + H_2O \tag{2}$$

$$4PbO + H_2SO_4 \leftrightarrow 3PbO \cdot PbSO_4 \cdot H_2O \tag{3}$$

On the other hand, two different boundary conditions are established regarding the temperature. If the mass temperature is higher than 75°C, most of the mass is composed of 4BS crystals along with α and β PbO Eqs. (4, 5). Nevertheless, the composition is affected when the temperature is below 60°C since the mass is mainly composed of 3BS, α PbO, and β PbO crystals [2, 9].

$$5PbO + H_2SO_4 \leftrightarrow 4PbO \cdot PbSO_4 + H_2O \tag{4}$$

$$2(3PbO \cdot PbSO_4 \cdot H_2O) + \alpha\ PbO + \beta\ PbO \rightarrow 2(4PbO \cdot PbSO_4) + 2H_2O \tag{5}$$

Finally, as it was previously mentioned, the composition of the active mass is influenced by the H_2SO_4/PbO ratio. Even the size of the 3BS crystals is affected by this6factor since it is reduced due to the increment of the ratio. So at H_2SO_4 / PbO ratio < 4%, the 3BS formation is developed and partially inhibited by a parallel mechanism caused by the significant alkaline conditions. Thus, the mix of PbO species (α PbO + β PbO) is hydrated Eq. (6) and consequently, the β PbO level is increased Eq. (7). Hence, the content of the 3BS precursor (α PbO) is reduced. At a ratio of 4% < H_2SO_4 / PbO < 8%, the formation of 3BS is less inhibited, but the unreacted PbO content remains significant. Furthermore, the hydrated secondary species are affected by the higher H_2SO_4 concentration. Thus, new amorphous species are formed Eq. (8). On the other hand, when the ratio is increased (8% < H_2SO_4/ PbO ratio < 10%), the paste is mainly composed of 3BS crystals, although there are others, such as 1BS and β PbO. However, if the PbO is previously mixed with water, the presence of β PbO is blocked. Therefore, under these conditions, the formation of 1BS is achieved, and the pH influence on β PbO is confirmed. Even so, at H_2SO_4/PbO ratio of around 5%, the formation of 4BS crystals is favored, and the 3BS levels are reduced [2, 8].

$$PbO + H_2O \rightarrow Pb(OH)_2 \tag{6}$$

$$Pb(OH)_2 \rightarrow \beta\, PbO + H_2O \tag{7}$$

$$2Pb(OH)_2 + H_2SO_4 \rightarrow Pb(OH)_2 \cdot PbSO_4 + 2H_2O \tag{8}$$

As explained at the beginning of this section, there are other amorphous species in the paste, such as hydroxides, hydrated lead sulfates, and carbonates. The presence of these compounds (15% of the paste content) is considered important because the lead sulfate crystals can be bonded to each other through them. Hydroxides and hydrated lead sulfates are generated by the hydration of PbO and 3BS compounds. In the case of PbO, part of its content is previously generated by a portion of free Pb (around 2.5%) in the early stages of massing Eq. (9). In this way, when the temperature is increased during the procedure and 3BS formation is carried out, the hydration of PbO and newly formed 3BS crystals Eqs. (6, 10) is performed. Furthermore, if the massing time is longed and the presence of CO_2 is increased, lead hydrocarbonate compounds will be also formed Eq. (11) [2].

$$2Pb + O_2 \rightarrow 2PbO \tag{9}$$

$$3PbO \cdot PbSO_4 \cdot H_2O + 2H_2O \rightarrow 3Pb(OH)_2 \cdot PbSO_4 \tag{10}$$

$$10Pb(OH)_2 + 6CO_2 \rightarrow PbO \cdot 6PbCO_3 \cdot 3Pb(OH)_2 + 7H_2O \tag{11}$$

4. Curing plate

Once the massing of the active material is finished and the subsequent grid pasting is performed, the new plates can be cured. The curing procedure is carried out in chambers and is usually divided into two stages. In the first one, which is properly considered as the curing step, the initial moisture of the active material must be exceeded by the set relative humidity (RH). After hours of curing, the later drying step is performed. In this case, the RH must be reduced below the moisture of the active material. In addition, the temperature can be adjusted to different values during the curing process. Plates rich in 3BS crystals will be obtained if the temperature is limited to 60°C. Thus, crystals up to 2 μm in length will be obtained and they will be interconnected, building agglomerates. Even so, if the aim is focused on the formation of 4BS crystals, the temperature will be increased above 80°C. Regardless of whether the active material has been previously produced with 3BS or 4BS compounds. Therefore, crystals up to 20 μm in length and 5 μm in diameter with a good interconnection between themselves will be found [2, 10].

For this reason, according to the above explanation, the 3BS crystals of pasted plate can be modified to 4BS crystals by a curing procedure. This mechanism is firstly initiated by the reaction of the unreacted α PbO particles to the β PbO polymorphic ones Eq. (12). And meanwhile, the 3BS crystals and unreacted PbO particles are hydrated Eqs. (6, 13). Later, a new formed hydrated compound of 3BS and PbO is observed. As a result, a complex hydrated structure of 4BS crystal is generated Eq. (14). Finally, 4BS crystal is obtained by dehydration of the previous complex structure Eq. (15) [2, 11, 12].

$$\alpha\, PbO \leftrightarrow \beta\, PbO \tag{12}$$

$$3PbO \cdot PbSO_4 \cdot H_2O + aH_2O \rightarrow 3Pb(OH)_2 \cdot PbSO_4 \cdot bH_2O \tag{13}$$

$$\begin{aligned}&\left[3Pb(OH)_2 \cdot PbSO_4 \cdot bH_2O + Pb(OH)_2 + cH_2O\right] \\ &\rightarrow \left[4Pb(OH)_2 \cdot PbSO_4 \cdot dH_2O + aH_2O\right]\end{aligned} \tag{14}$$

$$\left[4Pb(OH)_2 \cdot PbSO_4 \cdot dH_2O + H_2O\right] \rightarrow 4PbO \cdot PbSO_4 \cdot eH_2O + bH_2O \tag{15}$$

On the one hand, a hard porous internal mass (or "skeleton") is built up in the active material after the curing procedure. The mechanism is generated by the evaporation of the water portion. As a consequence, an interconnection between the different crystals and the unreacted PbO particles is generated. As it is showed in the **Figure 1**, the water portion indicated by the orange section is evaporated during the curing process. Thus, a hard porous mass structure is represented as a result. Furthermore, as long as the moisture of the active material is fixed at around 9%, the amorphous compounds solved and lead sulfate crystals will be recrystallized, causing a growth of their sizes. Lastly, while all these changes are carried out, the remaining free Pb content is oxidized to PbO Eq. (9). On the other hand, the performance of a positive will be influenced by the curing procedure. Since the basis of its structure is generated after this manufacturing step. Instead, the negative plate will not be affected at all because a new internal structure is generated in the following formation step.

After hours of the first curing stage, the mass structure is mechanically weak. For this reason, the active material moisture will be reduced to 0.2% in the subsequent drying stage. In this way, part of the water content in the capillaries and between the particles is evaporated in the beginning. Consequently, internal pores are formed and

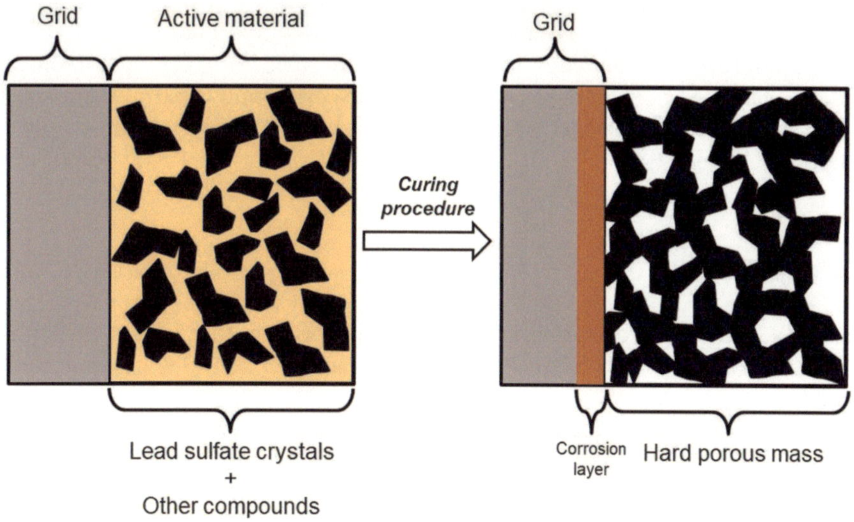

Figure 1.
Changes caused in the plate by the curing procedure.

the active material is shrunk. Later, while the drying step is extended, the residual water content is finally evaporated. Furthermore, the hydroxide species solved in this residual water content is precipitated and bound to the dried particles. Thus, a hard porous mass structure into the active material is achieved [2].

Apart from the prior internal mechanisms, there is another one in which the internal active material can be adhered to the grid (**Figure 1**). This mechanism is the formation of the corrosion layer on the grid surface. During the curing process, the layer is formed and its thickness is increased as well. As it is reflected in the **Figure 1**, a new "red" corrosion layer is formed between the grid and the active material. The mechanism is based on ion diffusion through oxygen vacancies. In this way, oxygen vacancies are formed with the oxidation of Pb at the interface between the grid and the active material Eq. (16). The electrons and oxygen vacancies then are moved from the formed corrosion layer to the interface between the layer and the active material. So, they are reacted by H_2O and O_2, obtaining lead hydroxides Eqs. (17–20). In addition, during the first curing stage, the decrease of water content must be compensated by a high RH value (more than 40%) for a good development in the mechanism [2].

$$Pb \rightarrow PbO_n^{2+} + 2e^- \tag{16}$$

$$PbO_n^{2+} + 2e^- + H_2O \rightarrow PbO + H_2 \tag{17}$$

$$PbO_n^{2+} + 2e^- + \frac{1}{2}O_2 \rightarrow PbO \tag{18}$$

$$PbO + H_2O \rightarrow Pb(OH)_2 \tag{19}$$

$$PbO_n^{2+} + 2e^- + \frac{1}{2}O_2 + H_2O \rightarrow Pb(OH)_2 \tag{20}$$

Talking more about the structure of the corrosion layer, two structural levels are found. The most internal one is basically constituted by PbO particles. These ones are previously obtained due to the oxidation of free Pb. The second level is obtained as a result of the hydration of PbO to $Pb(OH)_2$. Thus, the second level constituted by hydrated PbO particles can be bonded with the partially hydrated 3BS and 4BS crystals. Therefore, the connection is formed by the surfaces of the hydrated compounds. In the case of 3BS crystals, the lead sulfate compound is partially incorporated into the second level of the corrosion layer. Instead, the 4BS crystals are simply bonded to the layer [2].

5. Soaking plate

The first step of the formation is the soaking procedure. In this new stage, the cured plates are stored in H_2SO_4. In this way, the active material will be sulfated and hydrated.

The active material composition is considered as an important fact in the soaking procedure. Since the following mechanisms will be quite influenced by the content of either 3BS or 4BS crystals. For this reason, the procedure can be more accelerated with a 3BS plate. In addition, the soaking is influenced by the acid density. Therefore, at low-density values (1.06 g/mL), the structure of the positive active material (PAM) and negative active material (NAM) will be not significantly affected during the procedure. Even no distinctions between the soaking of 3BS or 4BS plates will be found. On the other hand, at higher values of sulfuric acid density (1.20 g/mL), the NAM and PAM structures will be modified during soaking. And the process on 3BS plates will be more significant [2].

Therefore, at low H_2SO_4 density values, the soaking procedure will be limited by the acid diffusion. Furthermore, the active material structure (especially the positive one) obtained after the subsequent formation will be defined by the pre-soaking procedures. Nevertheless, if the H_2SO_4 density is increased, the soaking will be limited by the active material reactivity. Additionally, the structure of the subsequently formed active material will be fixed by the soaking procedure [2, 11].

The soaking is carried out by a direct contact between the plate surface and the H_2SO_4 Eqs. (3, 21, 22), provided that the active material is composed of 3BS crystals and PbO particles. Thus, a $PbSO_4$ layer is formed on the plate surface. Additionally, part of this layer will be resolved and then recrystallized, generating larger $PbSO_4$ crystals. Later, when the soaking is prolonged, the plate porous structure will be penetrated by the acid. In this way, the 3BS crystals and PbO particles located in the pores are attacked by H_2SO_4, forming water, $PbSO_4$, and 1BS crystals Eqs. (3, 21, 22). As a consequence, the acidic solution will be modified to neutral one inside the plate. So, as soon as the new pores zones are reached by this solution, the formation of PbO, 1BS, and 3BS hydrated is executed Eqs. (6, 23). Then, the new hydrated species are sulfated by the acid flow Eq. (24). And even part of these sulfated species can be dehydrated Eq. (25) or sulfated again Eq. (26). Therefore, some areas of the inner of the plate will be composed of crystals and amorphous compounds. Furthermore, while the active material is penetrated by the acid, the growth of the $PbSO_4$ layer is extended to the plate surface. As a result, the sulfation process is slowed down because the penetration of H_2SO_4 is impeded. Since the size of the newly formed $PbSO_4$ crystals is increased on the plate surface. Finally, after more than 4 h of soaking, most of the 3BS crystals and PbO particles will be hydrated in the deepest zone of the plate [2].

$$3PbO \cdot PbSO_4 \cdot H_2O + H_2SO_4 \leftrightarrow 2(PbO \cdot PbSO_4) + 2H_2O \tag{21}$$

$$PbO \cdot PbSO_4 + H_2SO_4 \leftrightarrow 2PbSO_4 + H_2O \tag{22}$$

$$3PbO \cdot PbSO_4 \cdot H_2O + 3H_2O \rightarrow 3Pb(OH)_2 \cdot PbSO_4 \cdot H_2O \tag{23}$$

$$3Pb(OH)_2 \cdot PbSO_4 \cdot H_2O + H_2SO_4 \rightarrow 2\left(Pb(OH)_2 \cdot PbSO_4\right) + 3H_2O \tag{24}$$

$$Pb(OH)_2 \cdot PbSO_4 \rightarrow PbO \cdot PbSO_4 + H_2O \tag{25}$$

$$Pb(OH)_2 \cdot PbSO_4 + H_2SO_4 \rightarrow 2PbSO_4 + 2H_2O \tag{26}$$

The soaking process will be initiated in the same way if the plate is basically conformed by 4BS crystals. Thus, a $PbSO_4$ layer is formed on the plate surface due to the direct contact between the 4BS crystals and the H_2SO_4 Eq. (27). Nevertheless, as the 4BS crystal size is remarkable, only its surface layer will be completely transformed into $PbSO_4$. In addition, the superficial $PbSO_4$ crystals will be resolved and later recrystallized, generating larger crystals. When the soaking is prolonged, the active area in which the plate can be penetrated by the acid is reduced. So, the acid penetration into the plate will be more and more impeded. Hence, the pH value of the acid solution located in the pores will be increased. Consequently, the formation of 1BS and 3BS crystals will be carried out. Therefore, the inside of the plate will be composed of $PbSO_4$ and other lead sulfate crystals [2, 11].

$$4PbO \cdot PbSO_4 + 4H_2SO_4 \leftrightarrow 5PbSO_4 + 4H_2O \tag{27}$$

During the soaking procedure, the corrosion grid layer is reached by the acid flow and consequently, $PbSO_4$ crystals are formed. Even so, the inner of the grid will not be affected due to the acid will be isolated by the $PbSO_4$ layer recently generated. Additionally, as the acid flow is diluted into the active material pores, the interface between the active material and the corrosion layer will be hydrated as well. This mechanism is similar to plates rich in 3BS or 4BS crystals.

6. Formation of the active material

In this new manufacturing step, the active non-conductive material of the plates is transformed into an electrically conductive element. In this way, the initial compounds such as PbO particles, $PbSO_4$ and basic lead sulfate crystals are reduced to Pb or oxidized to PbO_2 on the negative and positive plate respectively. So, electromotive forces can be generated by the LAB with formed plates mechanically connected.

6.1 Positive plate

With the first current flow, PbO_2 is formed in the grid because it is the only electronic conductor element on the plate. Then, the formation of the active material is developed through two well-defined stages. In the first one, the α PbO_2 crystal structure is mainly obtained in the inner of the plate. Because the PbO particles and the basic lead sulfate crystals are oxidized Eqs. (28–30). In the case of 3BS crystals, the process is begun with its hydration Eq. (31). Later, the oxidation state of lead is increased Eq. (32) and followed by the formation of α PbO_2 Eq. (33). In a parallel, the formation of $PbSO_4$ Eq. (34) is detected from the previous hydration Eq. (31). Thus, the internal pores of the plates are also sulfated [13].

$$PbO + H_2O \rightarrow PbO_2 + 2H^+ + 2e^- \tag{28}$$

$$PbO \cdot PbSO_4 + 3H_2O \rightarrow 2PbO_2 + SO_4^{2-} + 6H^+ + 4e^- \tag{29}$$

$$3PbO \cdot PbSO_4 \cdot H_2O + 4H_2O \rightarrow 4PbO_2 + SO_4^{2-} + 10H^+ + 8e^- \tag{30}$$

$$3PbO \cdot PbSO_4 \cdot H_2O + 2H_2O \rightarrow 4Pb^{2+} + SO_4^{2-} + 6OH^- \tag{31}$$

$$Pb^{2+} \rightarrow Pb^{4+} + 2e^- \tag{32}$$

$$Pb^{4+} + 2H_2O \rightarrow PbO_2 + 4H^+ \tag{33}$$

$$Pb^{2+} + SO_4^{2-} \rightarrow PbSO_4 \tag{34}$$

Even so, this first stage is not only located in the core of the plate because other processes are found near the plate surface. In these reactions, the PbO and 3BS compounds are sulfated Eqs. (21, 22), obtaining1BS and $PbSO_4$ crystals as products. Because of these separate mechanisms, different structures are found in the formed PAM.

The second stage is achieved when the potential is increased to about 200 mV and the formation is carried out for at least 6 h. Then, the required oxidation potential of $PbSO_4$ is reached, and therefore, the formation of the other PbO_2 crystal structure (β PbO_2) is developed Eq. (35) [13].

$$PbSO_4 + 2H_2O \rightarrow PbO_2 + SO_4^{2-} + 4H^+ + 2e^- \quad (35)$$

Normally, the distribution of the two PbO_2 phases is well established. On the one hand, the α phase is located inside the plate, and the β phase is usually found in outer areas. Furthermore, as it was demonstrated in the prior mechanisms, the flow of H_2O, H^+ and SO_4^{2-} ions are crucial for the positive plate formation. Since the pH of the active material is compensated by the crossover of ions and water molecules between the active material and the electrolyte [13].

In contrast, the formation mechanisms are altered if the positive plate is rich in 4BS crystals. When the period of formation is longer than 15 h, a higher percentage to 70% is not reached in the formed plate. This low formation rate is caused by the complexity of the mechanism. Firstly, the external surface of $PbSO_4$ crystals is oxidized to PbO_2 Eq. (35), generating H_2SO_4 as a product. Thus, the internal 4BS crystals are partially reacted by the newly formed acid Eq. (27). So, an innermost surface of 4BS crystal will be composed of a 4BS + $PbSO_4$ layer. Thus, this new layer will be coated by PbO_2. Lastly, the inner of the 4BS crystal will be oxidized to PbO_2 provided that the active material is penetrated by H_2O. And the H^+ andSO_4^{2-} ions are driven out to the electrolyte. Nevertheless, the mechanism can be impeded due to the low ion transport rate. Therefore, the formation process is further impeded when the plate is composed mainly of 4BS crystals [13].

Lastly, the corrosion layer formed on the grid surface is oxidized during the forming process. There are two oxidation ways according to the dissolution of the external $PbSO_4$ layer. The mechanism without dissolution is started by the partial oxidation of $PbSO_4$ crystals Eq. (35). Thus, the surface of the corrosion layer is composed of $PbSO_4$ + PbO_2. Nevertheless, this layer will be completely oxidized during the battery cycling life. In the case of a dissolution mechanism, the corrosion layer is first oxidized to PbO_2 Eq. (35). Later, the $PbSO_4$ crystals are solved and the Pb^{2+} ions are oxidized near the new PbO_2 layer, forming a thicker layer. Lastly, the solved $PbSO_4$ crystals will be involved by a PbO_2 layer [2].

6.2 Negative plate

The formation of the negative plate starts on the grid. Afterward, the process is conducted on the plate surface. Firstly, the entire surface is reacted and then the process is aimed at the inner of the active material.

According to previous studies, the formation can be divided into two stages. In the first one, PbO particles and lead sulfate crystals are reduced to Pb Eqs. (36–38). But PbO and 3BS compounds are partially reduced because the rest is attacked by the H_2SO_4 Eqs. (21, 22). Therefore, in a plate constituted by 3BS crystals, the mechanism is carried out by the reaction between lead sulfate crystals and H_2O Eq. (39). Then, the oxidation state of the obtained Pb^{2+} ions are transformed and reduced to Pb Eq. (40). Additionally, as parallel reactions, other products of 3BS hydration are generated (H_2O Eq. (41) and $PbSO_4$ crystals Eq. (34). In this way, the pH of the system is changed to near neutral [13].

$$PbO + 2e^- + 2H^+ \rightarrow Pb + H_2O \quad (36)$$

$$3PbO \cdot PbSO_4 \cdot H_2O + 8H^+ + 8e^- \rightarrow 4Pb + H_2SO_4 + 4H_2O \quad (37)$$

$$PbO \cdot PbSO_4 + 4H^+ + 4e^- \rightarrow 2Pb + H_2SO_4 + H_2O \quad (38)$$

$$3PbO \cdot PbSO_4 \cdot H_2O + 2H_2O \rightarrow 4Pb^{2+} + SO_4^{2-} + 6OH^- \quad (39)$$

$$Pb^{2+} + 2e^- \rightarrow Pb \quad (40)$$

$$OH^- + H^+ \rightarrow H_2O \quad (41)$$

After more than 6 h of formation process, when the inner of the active material is not completely reduced to Pb, the second stage is performed. In this new stage, the reduction of $PbSO_4$ crystals to Pb Eq. (42) is located in the outer area of the plate. Thus, when the first stage is ended, the $PbSO_4$ crystal reduction is aimed at the core of the plate. Thus, as soon as the second stage is finished, the pH value decreased because H_2SO_4 is obtained as a product [13].

$$PbSO_4 + 2H^+ + 2e^- \rightarrow Pb + H_2SO_4 \quad (42)$$

Finally, the rate of the negative plate formation is controlled by the solved ion diffusion. Because during the first stage, the reaction layer is overloaded by negative solved ions (OH− and SO_4^{2-}). In this way, as the electronic neutrality must be achieved, an ionic crossing is induced between the active material and the electrolyte. So, the plate is penetrated by H^+ ions from the electrolyte and the newly formed SO_4^{2-} ions are expelled from the same plate [13].

7. Battery performance

After the formation process, the battery can be considered as an operative energy storage system. In state of charge, the PAM is composed of two different PbO_2 crystalline structures: the orthorhombic PbO_2 (α PbO_2) and the tetragonal PbO_2 (β PbO_2). And the NAM is composed of Pb. In addition, the role of the electrolyte as reactant is observed as well (**Figure 2**). Therefore, in the discharge process (see **Figure 2**), PAM will be reduced to $PbSO_4$ and water will be formed as a product Eq. (43). And NAM will be oxidized to $PbSO_4$, and proton ions will be solved into the electrolyte Eq. (44). Thus, a voltage of 2.0 V is provided by both plates Eq. (45) [7].

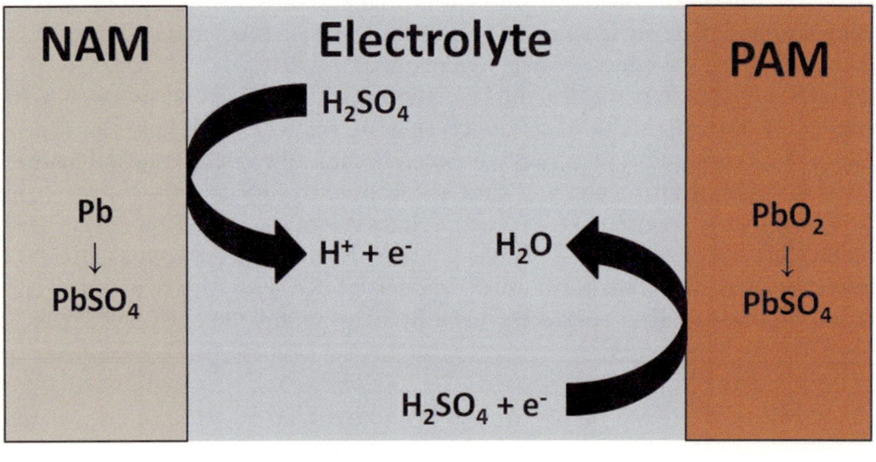

Figure 2.
A scheme of a LAB discharge process.

$$PbO_2 + H_2SO_4 + 2H^+ + 2e^- \leftrightarrow PbSO_4 + 2H_2O \quad (43)$$

$$Pb + H_2SO_4 \leftrightarrow PbSO_4 + 2H^+ + 2e^- \quad (44)$$

$$PbO_2 + Pb + 2H_2SO_4 \leftrightarrow 2PbSO_4 + 2H_2O \quad (45)$$

During the charging process, PAM will be oxidized to PbO_2 Eq. (43) and the NAM will be reduced to Pb Eq. (44). However, the charge current is consumed by other secondary reactions due to their enhanced kinetics. These reactions are located on the surface of the positive and negative plates. Therefore, on the PAM`s surface, the oxygen evolution reaction (OER) is observed Eq. (46), and on the NAM`s surface, the hydrogen evolution reaction (HER) is found Eq. (47). So, the battery will be negatively affected by a high water consumption and an increase of its internal resistance if the HER and OER are not controlled [14]. Besides, part of the O_2 generated on PAM can be driven to the NAM surface and therefore reduced to H_2O Eq. (48), this mechanism is known as "recombination process."

$$H_2O \rightarrow \frac{1}{2}O_2 + 2H^+ + 2e^- \quad (46)$$

$$2H^+ + 2e^- \rightarrow H_2 \quad (47)$$

$$\frac{1}{2}O_2 + 2H^+ + 2e^- \rightarrow H_2O \quad (48)$$

One of the most common issues during the LAB operation is the plate sulfation. This issue is caused by the irreversibility of the $PbSO_4$ crystals formed after the discharge process. Focusing on the negative plate, after numerous charge and discharge cycles, some of the $PbSO_4$ crystals cannot be reduced to Pb. Additionally, the crystal size is increased in each cycle. In this way, a battery failure can be caused due to this mechanism. Therefore, the harmful effects of the sulfation can be reduced by the use of additives. For example, the incorporation of expander additives is experimentally confirmed as a useful strategy. Furthermore, other materials, such as organic expanders, the $BaSO_4$, and the carbon compounds are also used to prevent sulfation. In the case of carbon additives, different options can be used: the carbon black, the graphite, the graphene, and the activated carbons. Therefore, longer battery cycle life can be achieved through an appropriate content of the above additives into the NAM [15].

In the **Figure 3**, cured plates of PAM (A) and NAM (B) are showed. In this case, the plates were assembled in a flooded battery for the transportation market. Besides, a commercial battery (C) and its system (D) are showed as example.

8. Premature capacity loss: Corrosion layer growth in the positive plate

When a LAB is overcharged, the main reactions on the plates are moved to a secondary role because most of the $PbSO_4$ crystals are reacted to Pb (negative plate) or PbO_2 (positive plate). In this way, as it was previously explained in the section 7, the charge current is consumed by other secondary reactions, on the negative plate: the HER and ORR and on the positive plate: the OER. Nevertheless, the grid corrosion of the positive plate is found as another important secondary reaction [16].

Deeping into this process, this mechanism originated when the LAB is charged or overcharged. As it is represented in the **Figure 4**, the process is started when the

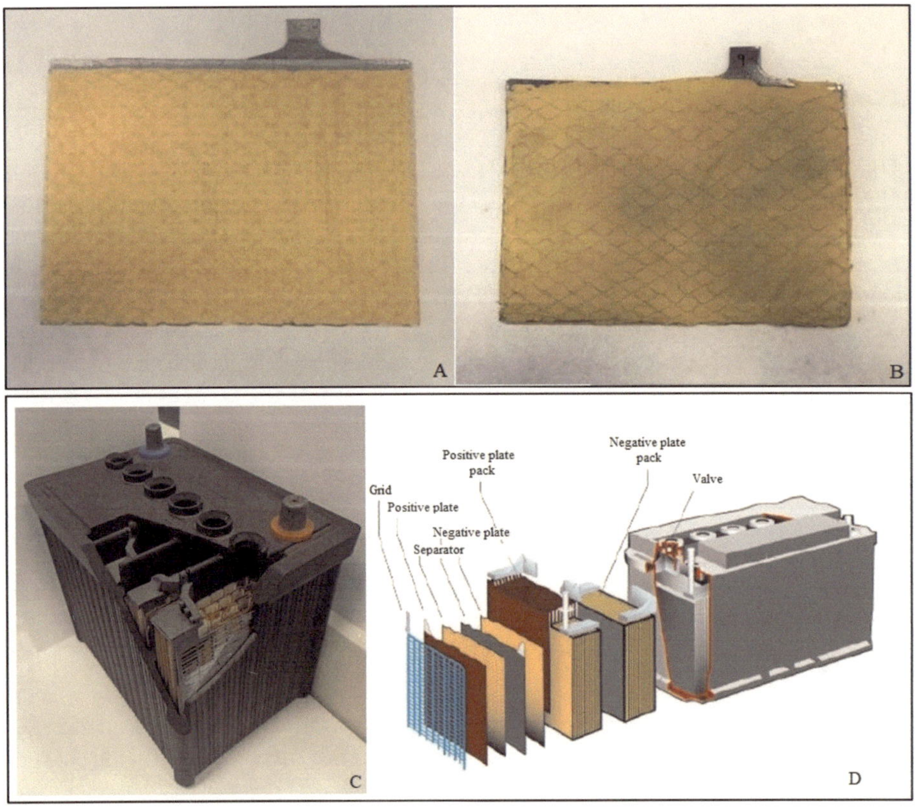

Figure 3.
A) Example of a cured PAM. B) Example of a cured NAM. C) Example of a commercial automotive LAB. D) Scheme of the LAB assembly.

PAM and the corrosion layer are penetrated by the oxygen generated in the OER. The sense of the penetration is indicated by the arrows in the **Figure 4**. Thus, the surface of the grid is attacked by oxygen and oxidized to PbO. Then, the formed PbO may then undergo further oxidation, generating PbO_x species. This new layer is represented by the pink color in the **Figure 4**, and it is located between the active material and the corrosion layer interface. In addition, the process is favored by two facts: the high anodic potential values reached on the positive plate and its thermodynamic nature. Since the process is spontaneous with the contact between the PAM and the grid. Even so, the kinetic of the process may be reduced by the passive layer formed between the active material and the grid [17].

Lastly, the battery performance can be adversely affected by the corrosion process if the PbO reaction rate is more favored than the PbO_x one Eq. (49). Consequently, an ohmic resistance layer would be formed causing the polarization of the positive plate, the gradual loss of capacity and the formation of pores and cracks in the grid. This issue was avoided by different studies focused on the alloy composition of the positive grid [2, 13, 16–19].

$$Pb + \frac{x}{2}O_2 \rightarrow PbO_x \tag{49}$$

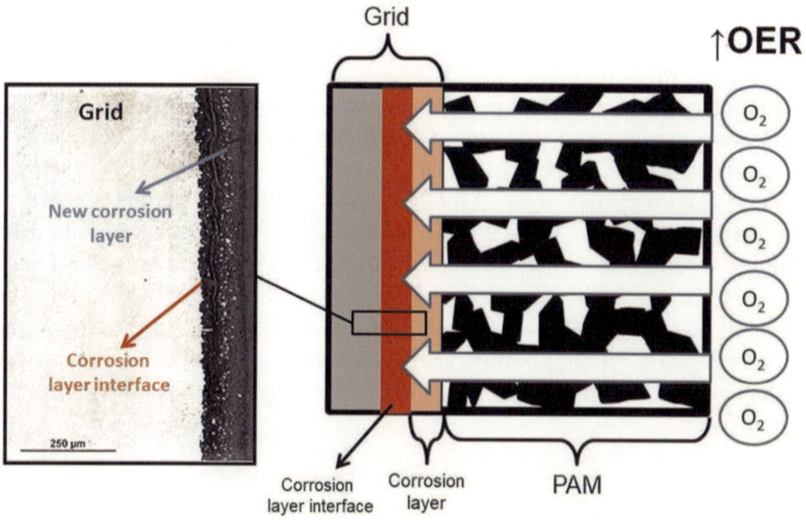

Figure 4.
Real example of a corrosion layer growth and a representation of its mechanism.

9. Battery improvements: Studies focused on the active material as a new research approach

According to the important role of the active material, different research projects were carried out to improve the LAB performance through enhancements of the internal structure. One research strategy was focused on the synthesis of a more porous and spongier nanostructured PbO. For this purpose, Karami et al. applied ultrasonic waves with fixed frequencies as an experimental synthesis tool [3]. Thus, they obtained PbO after using $Pb(NO_3)_2$ and NaOH as initial reagents and applying different sonication, dehydration, drying, and filtration procedures. Then, they used the new nanostructured PbO to build negative and positive plates. As a result, better capacity values and higher number of life cycles were showed by the batteries fabricated with the new PbO [3].

The use of a LAB at high-rate partial state of charge (HRPSoC) is increasing every year. However, the performance of the plates at these operation modes can be improved. Since the sulfation process is very aggressive during its performance. Because of this issue, improvements to the internal porous structure of the active material or the use of common battery additives with new structural or chemical modifications have been studied.

On the one hand, Jin et al. studied how the control of the PbO_2 particle size of the positive plate can extend the battery life since the reversibility of $PbSO_4$ crystals could be improved [4]. In this way, they compared two formed PAMs whose PbO_2 particle sizes were 10 and 100 nm in diameter. After 75 days of discharge, the batteries were recharged for 24 h. During these periods, the phase transformations of the PAM were studied through multiple techniques: scanning electron microscope (SEM), focused ion beam (FIB) combined with transmission electron microscopy (TEM) cross-section, and a thermal test. After the research, a longer lifetime was showed by the 100 nm PAM due to the shedding of its active material was reduced [4]. On the other hand, Yang et al. studied the use of PbO/graphene composites as an additive in NAM.

In this way, the aggregation of irreversible $PbSO_4$ crystals could be reduced and the reversibility of the Pb/$PbSO_4$ system could be improved. Thus, they synthesized the compound using graphene oxide and $Pb(NO_3)_2$ as reagents. Lastly, they obtained the final product after sonication and pyrolysis procedures. The results showed that part of the PbO was absorbed by the graphene oxide layers causing improvements in the battery cycle life and a reduction in the HER rate [5].

One of the newest concepts in LAB technology is the ultra battery system. In this type of battery, the negative plate is covered by a carbon layer or the plate itself is compressed to a capacitor electrode. Thus, a higher charge acceptance and longer cycle life will be achieved by this new technology. In keeping with the above information, Thangarasu et al. measured the effect of a Pb nanoparticle impregnated negative carbon plate. Finally, better results were found in the cell capacity, the HER rate and the cycle life through the use of this negative plate covered by a modified carbon layer [6].

10. Conclusions

As demonstrated, the LAB performance is highly dependent on the manufacturing process. Since the structure and composition of the active material are modified during the distinct production steps. These changes can be summarized by two important ones: the formation of the hard porous mass and the corrosion layer in the curing process. And the oxidation and reduction of the active material during the formation procedure. Furthermore, the previous phase transformations are highly influenced by the manufacturing conditions and some critical species. Because the battery could not be lead an electromotive force system without their important role. In the case of the species, the non-reacted PbO particles, the lead sulfate crystals, and the solved ions stand out from the rest. Moreover, these phase changes are not limited to the production process, as many others occur during the battery lifetime. For example, the sulfation of the plates during the discharge process or the secondary reactions occurring in the charging process, such as the water consumption, recombination, and corrosion layer growth. For this reason, recent researches have been aimed to improve the performance of the plates by structural modifications of their active material. In this way, the negative effects caused by sulfation or water consumption can be reduced, and therefore, the battery performance can be improved.

Nomenclature

LAB	Lead-acid battery
VRLA	Valve-regulated lead acid
α PbO	Polymorphic structures (tetragonal)
β PbO	Polymorphic structures (orthorhombic)
α PbO_2	Polymorphic structures (orthorhombic)
β PbO_2	Polymorphic structures (tetragonal)
1BS	Monobasic lead sulfate $PbO{\cdot}PbSO_4$
3BS	Tribasic lead sulfate $3PbO{\cdot}PbSO_4{\cdot}H_2O$
4BS	Tetrabasic lead sulfate $4PbO{\cdot}PbSO_4$
RH	Relative humidity
PAM	Positive active material

NAM	Negative active material
OER	Oxygen evolution reaction
HER	Hydrogen evolution reaction
HRPSoC	High-rate partial state of charge
SEM	Scanning electron microscope
TEM	Transmission electron microscopy
FIB	Focused ion beam

Author details

Alberto F. Romero[1,2]* and Pilar Ocón[2]

1 Exide Group, Azuqueca de Henares, Guadalajara, Spain

2 Department of Applied Physical Chemistry, Universidad Autónoma de Madrid, Madrid, Spain

*Address all correspondence to: alb.fco.romero.rodriguez@gmail.com

References

[1] May GJ, Davidson A, Monahov B. Lead batteries for utility energy storage: A review. Journal of Energy Storage. 2018;**15**:145-157

[2] Pavlov D. Lead-Acid Batteries: Science and Technology. 1st ed. Elsevier Science; 2011. pp. 253-479

[3] Karami H, Karimi MA, Haghdar S, Sadeghi A, Mir-Ghasemi R, Mahdi-Khani S. Synthesis of lead oxide nanoparticles by sonochemical method and its application as cathode and anode of lead-acid batteries. Materials Chemistry and Physics. 2008;**108**: 337-344

[4] Jin J, Jin D, Shim J, Shim W. Enhancing reversible sulfation of PbO_2 nanoparticles for extended lifetime in lead-acid Batteries. Journal of the Electrochemical Society. 2017;**164**(7): A1628-A1634

[5] Yang H, Qi K, Gong L, Liu W, Zaman S, Guo X, et al. Lead oxide enveloped in N-doped graphene oxide composites for enhanced high-rate partial-state-of-charge performance of lead-acid battery. ACS Sustainable Chemistry & Engineering. 2018;**6**: 11408-11413

[6] Thangarasu S, Palanisamy G, Roh S, Jung H. Nanoconfinement and interfacial effect of Pb nanoparticles into nanoporous carbon as a longer-lifespan negative electrode material for hybrid lead–carbon battery. ACS Sustainable Chemistry & Engineering. 2020;**8**: 8868-8879

[7] Rand DAJ, Moseley PT. Secondary batteries lead–acid systems. Encyclopedia of Electrochemical Power Sources. 2009;**2009**:550-575

[8] Pavlov D, Papazov G. Dependence of the properties of the lead-acid battery positive plate paste on the processes occurring during its production. Journal of Applied Electrochemistry. 1976;**6**: 339-345

[9] Iliev V, Pavlov D. The influence of PbO modification on the kinetics of the $4PbO \cdot PbSO_4$ lead-acid battery paste formation. Journal of Applied Electrochemistry. 1979;**9**:555-562

[10] Prout L. Aspects of lead-acid battery technology 3. Journal of Power Sources. 1993;**41**:185-193

[11] Wagner R. Secondary batteries lead-acid systems curing and formation. Encyclopedia of Electrochemical Power Sources. 2009;**2009**:677-692

[12] Laruelle S, Grugeon-Dewaele S, Torcheux L, Delahaye-Vidal A. The curing reaction study of the active material in the lead-acid battery. Journal of Power Sources. 1999;**77**: 83-89

[13] Rand DAJ, Moseley PT, Garche J, Parker CD. Valve-regulated Lead-acid Batteries. 1st ed. Elsevier Science; 2004. pp. 37-108

[14] Misra SS. Secondary batteries lead–acid systems. Encyclopedia of Electrochemical Power Sources. 2009; **2009**:764-778

[15] Blecua M, Romero AF, Ocon P, Fatas E, Valenciano J, Trinidad F. Improvement of the lead acid battery performance by the addition of graphitized carbon nanofibers together with a mix of organic expanders in the negative active material. Journal of Energy Storage. 2019;**23**:106-115

[16] Wang F, Hu C, Zhou M, Wang K, Lian J, Yan J, et al. Research progresses of cathodic hydrogen evolution in advanced lead–acid batteries. Scientific Bulletin. 2016;**61**:451-458

[17] Yang J, Hu C, Wang H, Yang K, Liu JB, Yan H. Review on the research of failure modes and mechanism for lead-acid batteries. International Journal of Energy Research. 2016;**41**:336-352

[18] Gelbke M, Mondoloni C. Flooded starting-lighting-ignition (SLI) and enhanced flooded batteries (EFBs): State-of-the-art. Future Automobiles. 2017;**2017**:149-184

[19] Lam LT, Lim OV, Haigh NP, Rand DAJ, Manders JE, Rice DM. Oxide for valve-regulated lead-acid batteries. Journal of Power Sources. 1998;**73**:36-46

Section 4

Phase Change Material for Cooling Application

Chapter 6

Application of Semi-Circular Micro-Channel Heat Sink with Phase Change Material for Cooling of Electronic Devices

Sunny Chandra, Om Prakash and Anupam Prakash

Abstract

In today's era demand of fast processing speed and superior performance of electronic devices creates the problem of huge amount of heat generation. About 100 W/cm^2 heat fluxes has been generating from advanced microprocessors of latest electronic devices and heat dissipation limited to 37 W/cm^2 for air cooling. Semi-circular micro-channel heat sink with Phase change materials still needs to be investigated for the cooling of latest electronic devices as combination of large heat transfer area within a small volume of heat sink and large latent heat released from PCM provides huge amount of heat transfer coefficient which is required to be achieved in the present work. Analytical study of micro-channel heat sink has been performed in the present work at a constant power supply of 95 W using PCM slurry at $\varphi = 0.25$ and $\varphi = 0.50$ for flow rate in the range of 75 ml/min to 300 ml/min. Heat transfer coefficient of slurry and Nu increases with increase in flow rate. At $\varphi = 0.25$ higher heat transfer coefficient and Nu has been observed than $\varphi = 0.50$ due to high viscous flow at higher concentration of PCM slurry in micro-channels.

Keywords: phase change material (PCM), slurry concentration (φ), semi-circular micro-channel, latent heat, flow rate, heat transfer coefficient

1. Introduction

In present scenario demand of high processing speed from processors and compact size of electronic devices increases the problem of high heat generation because high processing speed of processors increases the power consumption and compactness of electronic devices increases the required number of transistors per chips which ultimately increases the electrical power input. It can be clearly understood from the Moore's law **Figure 1** [1] that how vastly the number of transistors per chip increased after the application of modern compact processors. **Table 1** [2] Clearly indicates that with the development of advanced processors (Intel) having higher processing speed the thermal design power consumption (TDP) of processors increases resulting in higher heat generation. Heat fluxes (heat generated per unit surface area) have

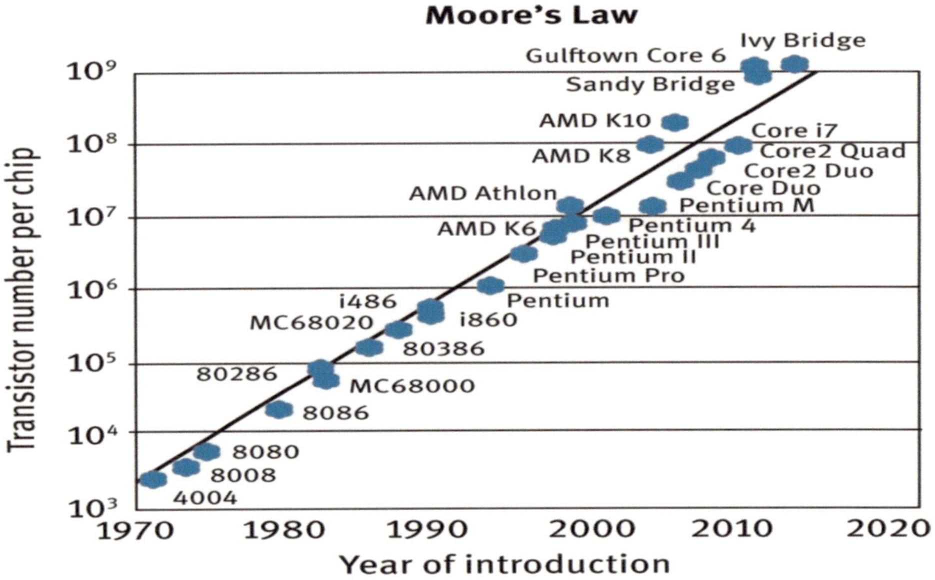

Figure 1.
Increase in number of transistors per chip with the development of processors over the year [1].

Model	Processing Speed (GHz)	TDP (W)
Core i7 860	2.8	95
Core i7 940	2.933	130
Core i5–750s	2.4	82
Core i5 2320	3.0	95
Core i3 530	2.93	73
Core i3 2100T	2.5	35
Core i3 550	3.2	73
Core 2 Duo E4300	1.8	65
Core 2 Quad Q6700	2.66	95
Core 2 Extreme QX6850	3.0	130
Pentium 4 1.3	1.3	51.6
Pentium 4 1.4 (Socket 423)	1.4	54.7
Pentium 4 2.26	2.26	56
Pentium III 450	0.450	33.76
Pentium III 500	0.500	37.52
Pentium III 550	0.550	39.8
Pentium II 233	0.233	34.8
Pentium II 266	0.266	38.2
Pentium II 300	0.300	43

Table 1.
Thermal design power (TDP) with processing speed of processors [2].

approached 100 W/cm^2 for advanced microprocessors while the heat fluxes dissipation limited to 37 W/cm^2 for air cooling [3], which ultimately increases the working temperature beyond the safe limit resulting in in-efficient performance and shorter life of the devices. The working temperature of electronic devices should be within 85°C-120°C to ensure the efficient performance of devices [4]. Hence an effective cooling method is needed for dissipating the excess heat flux generated. In this regard Micro-channel heat sink has emerged as a latest cooling method for controlling excess heat generated from a compact electronic device. Micro-channel heat sink provides high surface area to volume ratio for a given compact electronic device.

Usually Micro-channel heat sink is constructed from higher thermal conductivity materials such as copper, aluminum, brass etc. consisting channels of micron size (10 to 1000 μm) which acted as a passage of cooling fluid [5]. Only standard cooling technology would not be enough when electronic devices are continuously operated over longer periods due to space limitations and some limitations of thermal conductivity of cooling fluid and materials of heat sink the heat load generated cannot be properly dissipated. Phase change materials (PCM) can be utilized to take up excess heat and dissipate it from the electronic device ultimately improving the performance of an electronic device. PCM generally absorbs heat between the liquid and solid state. So micro-channel heat sink with PCM would be suitable to meet these requirements of heat flux dissipation also when PCM is applied in micro-channel heat sink thermal performance of micro-channel heat sink will enhance. Some literatures are available on the application of PCM in micro-channel heat sink for dissipating heat fluxes from electronic devices. Debich et al. [4] numerically investigated the heat sink based on PCM to find best configuration of heat sink. Analysis is established on the comparison of experimental results of heat sink with and without PCM. Effects of various parameters like geometry, boundary conditions, and materials have been considered for the numerical investigation. Results revealed that heat sink with n- Eicosane have better heat transfer compared with other PCMs based heat sink. Latent heating phase delayed due to increase in the volume fraction of PCM also when input power increases then melting rate of PCM increases ultimately increasing the thermal performance of heat sink. Hasan and Tabena [6] numerically explored the application of PCM in micro-channel heat sink utilizing air as a coolant. Four PCM (paraffin wax, neicosane, p116 and RT41) have been utilized as cooling mediums after using air as cooling fluid in different configurations of micro-channel heat sink at different ambient temperatures. Uniform heat flux is applied at the bottom of heat sink and mixed convection with radiation boundary is applied at the top surfaces. Results show that application of PCM in micro-channel heat sink with different configurations enhances the performance of micro-channel heat sink. The result also indicate that different PCMs cause unusual reduction in micro-channel heat sink surface temperature in the range of ambient temperature because of difference in melting point of different PCMs. Deng et al. [7] analytically investigated thermal performance of a micro-channel heat sink using microencapsulated PCM slurry for chip cooling. For analytical investigation of micro-channel heat sink fin and porous media methods have been used under constant heat flux condition. Significant differences in overall Nusselt number (Nu) and fluid temperature distribution have been predicted from the fin and porous media methods. Xu et al. [8] experimentally examined the heat transfer performance of a noble micro-channel heat sink consisted of a cavity with pyramid pin fins and a PCM cavity with square pin fins. Carbon nanotube (CNT) is added to paraffin to increase the thermal conductivity. The thermal performance of heat sink has been improved by using Graphene oxide particles (GOPs) nanofluids pulsating flow. Their results demonstrated that performance of heat sink improved significantly by utilizing PCM composite and GOPs nanofluids. The thermal resistance reduces significantly and the Nu surged up to 34.9% when the proportion of CNT was approximately 20%. The heat transfer performance has maximum increment under pulsating flow. The increased effect of exciting flow and PCM composite are also exaggerated by heating load and pump power. The higher heating load gives augmentation effect while pump power gives the decrement effect. Chang et al. [9] explored the performance of a modified heat sink with cavity on top portion of the heat sink for the purpose of storing composite PCM. The cavity was crammed with PCM composite having CNTs of concentration (9% and 15%). Their outcomes revealed that the thermal conductivities of composite PCM's were increased by 291.6% and 537.5% for the mass fraction 9% and 15% of graphite particles. It has been also found that the application of composite PCM decreases the system's maximum temperature significantly compared to pure paraffin. Karaipekli et al. [10] investigated the heat transfer performance of expanded CNTs composite/perlite/paraffine PCM. Results showed that the thermal conductivity increases by 26, 60 and 113.3% respectively for the mass fraction of CNTs 0.3%, 0.5%, and 1%. Rostamian et al. [11] experimentally examined the performance of electronic board heat sink utilizing pure and microencapsulated PCMs. Different geometries of heat sink like Square with three fins and seven fins, circular with 12 fins have been used for investigation over a range of electrical power 5 to 18 W in two cases preset and pulsed power based on the board critical temperature of 80°C. Experimental findings indicated that square seven fins heat sink to be the best among all configurations. Operating time increases when volume fraction of PX52 increases and it has been found that the critical temperature for PX52 was less than the critical temperature for pure PCM under the same conditions because of higher rate of heat transfer. Siyabi et al. [12] experimentally studied the consequences of PCM combination, adjustment of PCMs in multiple heat sink, PCM thickness, melting point and magnitude of heat source on conduct of three different heat sink. It has been found that (i) PCM combination RT50–RT55 enhances the temperature regulation period and also decreases the temperature of at the end point of the operation, (ii) the arrangement of RT58–RT47 decreases slightly the peak temperature compared to RT47–RT58 arrangement (iii) when thickness of PCM increased from 30 mm to 60 mm, the temperature regulation period increased by 50 min, (iv) when the melting point of PCM increases then the temperature regulation period and the heat sink temperature increases and (v) As the power rating increases from 1 to 2 W the temperature regulation period decreases. Kothari et al. [13] experimentally studied the competent thermal management system based on PCM to cool portable electronic devices. Paraffin wax was used as PCM for four different configurations of heat sink made of aluminum (unfinned heat sink with pure PCM, two finned heat sink with pure PCM, unfinned heat sink with MF-PCM composite and two finned heat sink with MF-PCM composite) in the present study to boost the operating time of heat sink to attain a critical set point temperature. Results showed that higher enhancement ratio and effective thermal control obtained with two finned heat sink with MF-PCM as compared to other configurations of heat sink. Thomas et al. [14] numerically investigate the heat transfer performance of a transferable electronic device by application of PCM in heat sink. n-eicosane with a melting point of 36.5°C has been used as PCM. Results indicated that for the increase in ambient temperature and decrease in latent heat PCM was melting as noted by increase in the power consumption. Gaikwad and More [15] experimentally investigated the potential of application of PCM slurry flowing through rectangular micro-channel heat sink for cooling of computer processor. Comparative analysis has been done for two working fluids water and PCM slurry with two different aspect ratios of 2 and 3 at flow rates ranging from 75 ml/min to 300 ml/min. The results show that cooling systems using PCM are advantageous not only over the conventional air cooling system but also on the water cooling system. Experimental results showed that lower maximum temperature of processor can be achieved at the same mass flow rate and pumping power by using PCM slurry over the water. Hence micro-channel heat sink with PCM slurries found to be more effective fluid in comparison to water as a cooling fluid. Kondle et al. [16] numerically investigated the thermal performance of PCM fluid for laminar flow conditions in rectangular and circular microchannel. Investigation has been performed under different aspect ratios and boundary conditions. Results showed that Nusselt number (Nu) increases when PCM slurry has been used as cooling fluid as compared to only water as cooling fluid for all aspect ratios. Kuravi et al. [17] numerically studied the performance of nano-encapsulated PCM slurry in manifold microchannel heat sink. Effect of different parameters has been considered for the study such as inlet temperature of fluid, concentration of PCM, and heat flux. Results showed that Nu increased for PCM slurry and bulk mean temperature of fluid reduces for PCM slurry as compared to pure single phase fluid. After studying above literatures it can be clearly concluded that thermal performance of a micro-channel heat sink can be enhanced by using PCM as slurry to flow in micro-channels or to adhere PCM in the heat sink. From the literatures it has been inferred that application of PCMs in a semi-circular micro-channel heat sink is yet to be studied. In this paper application of PCM as slurry has been studied using water as a base fluid in a micro-channel heat sink consisting of micro-channels with semi-circular shape at the end i.e. semi-circular micro-channels. Semi-circular micro-channels have been used because semi-circular shape of micro-channels can be easily created by wire EDM machining also this wire EDM is most economical among other methods of machining.

2. Selection of PCM for the study

PCM is a substance which has high latent heat of fusion i.e. melting and solidified at a certain temperature. It has ability to store and release large quantities of heat. PCM continues to absorb heat till all the materials are melted without significant rise in temperature. Hence PCM has been selected based on its melting temperature and accordingly maximum temperature rise of the heat sink. PCM can be classified as Organic, Inorganic and biomaterial etc. The cost of PCM increases with it's purity. Pure paraffin wax is more costly than technical grade paraffin. The cost of PCM is not the criteria for the selection of PCM in this study as the volume of the micro-channel heat sink is very low so very less amount of PCM will be required. In the present study PCM slurry of RT41 with water as base fluid has been used to flow through the semi-circular micro-channels.

3. Methodology

The model of micro-channel heat sink considered for analytical study is square cross-section of dimension 25 mm × 25 mm. The micro-channel heat sink consists of semi-circular micro-channels of 350 μm (0.35 mm) diameter with 0.20 mm of space between two channels as shown in **Figure 2**. The model has been studied for uniform

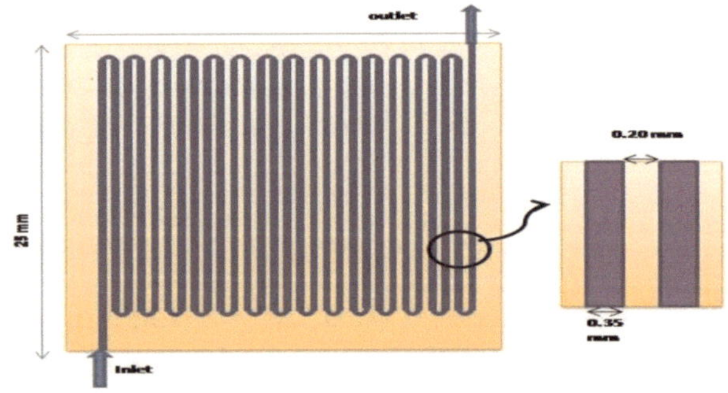

Figure 2.
Schematic diagram of semi-circular micro-channel heat sink.

Material	Thermal conductivity (W/mK)	Sp. heat (kJ/kgK)	Density (kg/m^3)	Temperature (°C)		Melting heat (kJ/kg)	Viscosity for liquid phase (Pa-s)
				Ts (solidus)	T_L (Liquidus)		
RT 41 (PCM)	0.2	2	802	37.5	42.9	141.7	0.003
Water	0.614	4.187	995.65	100		125.75	0.000798

Table 2.
Properties of fluid materials considered for the study [6].

power supply of 95 W from a heater attached to the bottom of the heat sink. In order to study the model of heat sink some assumptions have been considered.

- PCM slurry of RT 41 and water at inlet temperature of 30°C used as the working fluid. The properties of RT 41 and water are shown in **Table 2**.
- Flow rate of working fluid is varied from 50 ml/min to 100 ml/min for the volumetric concentration of 0.01, 0.25, and 0.50.
- Constant power of 95 W has been supplied at the bottom of the heat sink.

3.1 Properties of working fluid

Properties of working fluid (PCM slurry) have been calculated from the relations available in the literature [18]. The calculated properties of PCM slurry used as working fluid summarized below in **Table 3**.

Density (ρf)

$$\rho_{slurry} = \varphi \rho_w + (1 - \varphi)\rho_{PCM} \tag{1}$$

where φ is volumetric concentration of working fluid, ρw is density of water and ρPCM is the density of phase change material.

Viscosity (μf)

$$\mu_{slurry} = \left(1 - \varphi - 1.16\varphi^2\right)^{-2.5} \tag{2}$$

S. No.	Properties	Volumetric concentration (φ)	Value
1.	φ_{slurry} (kg/m³)	0.01	995.456
		0.25	947.237
		0.50	898.825
2.	μ_{slurry} (Pa-s)	0.01	8.183 × 10–4
		0.25	2.111 × 10–3
		0.50	3.947 × 10–2
3.	k_{slurry} (W/mK)	0.01	0.612
		0.25	0.489
		0.50	0.380
4.	c_{slurry} (kJ/kgK)	0.01	4.428
		0.25	10.360
		0.50	16.320

Table 3.
Calculated properties of working fluid.

Thermal conductivity (kf)

$$k_{slurry} = k_w \frac{2 + \left(k_{PCM}/k_w\right) + 2\varphi\left\{\left(k_{PCM}/k_w\right) - 1\right\}}{2 + \left(k_{PCM}/k_w\right) - \varphi\left\{\left(k_{PCM}/k_w\right) - 1\right\}} \tag{3}$$

where kw is the thermal conductivity of water and kPCM is the thermal conductivity of phase change material.

Specific heat (cf)

$$c_{slurry} = \frac{\varphi\left(c_{sp} + c_{Lp}\right)}{2} + \frac{\varphi \text{Latent heat}}{T_L - T_s} + \frac{(1-\varphi)\rho_w c_w}{\rho_f} \tag{4}$$

where c_{PCM} is the specific heat of phase change material and c_w is the specific heat of water.

3.2 Data interpretation

The working fluid used to flow as coolant through the semicircular micro-channels of 0.35 mm over the entire compact micro-channel heat sink of 25 mm² considered for the present work. The calculated properties of working fluid (PCM slurry) have been used for the analysis of the heat sink. Data assumed for the analysis of present work is summarized below in **Table 4**.

Power supply to the heat sink (P) in Watt

$$P = Q \times \rho_{slurry} \times c_{slurry}\left(T_{out} - T_{in}\right)$$

$$T_{out} = T_{in} + \frac{P}{Q \times \rho_{slurry} \times c_{slurry}} \tag{5}$$

Cross-section of heat sink (mm^2)	Diameter (D) of micro-channel (mm)	Length of single micro-channel (mm)	Space between two consecutive channels (mm)	Power Supply (W)	Working fluid	Flow rate of working fluid (ml/min)	Temperature at the inlet of heat sink (K)
Square (25 × 25)	0.35	15	0.2	95	Mixture of water and RT41 (PCM slurry)	75 to 300	303

Table 4.
Data assumption for the analytical analysis.

where Q is the flow rate of slurry in m3/s, Tout and Tin are outlet and inlet temperature of working fluid at the inlet and outlet of heat sink.

Reynolds Number (Re)

$$Re = \frac{\rho_{slurry} v D_h}{\mu_{slurry}} \tag{6}$$

Hydraulic diameter (Dh) can be calculated from

$$D_h = \frac{A}{P_w} \tag{7}$$

where A is the cross-sectional area of micro-channel and Pw is the wetted perimeter. For semi-circular cross-section

$$A = \frac{\pi D^2}{8} = 0.04810.mm^2 \tag{8}$$

So Dh for semi-circular micro-channels can be formulated as

$$D_h = \frac{\pi D}{\pi + 2} = 0.2138 \text{ mm} \tag{9}$$

Prandtl number (Pr)

$$Pr = \frac{c_{slurry} \mu_{slurry}}{k_{slurry}} \tag{10}$$

Nusselt Number (Nu) correlations with Re and Pr [19].

If 100 < Re < 1500 then $Nu = 0.253\ Re^{0.597}\ Pr^{0.349}$ and if 1500 < Re < 15,000 then

$$Nu = 0.253\ Re^{0.62}\ Pr^{0.340} \tag{11}$$

As we know that

$$Nu = \frac{h_{slurry} D_h}{k_{slurry}};\quad h_{slurry} = \frac{Nu \times k_{slurry}}{D_h} \tag{12}$$

Also heat absorbed (H) in Watt to the micro-channels is given by the relation

$$H = h_{slurry} A_{surface} (T_w - T_{in}) \tag{13}$$

where $A_{surface}$ is the surface area of the micro-channels and Tw is the wall temperature of the micro-channels.

Surface area of semi-circular micro-channels can be calculated as

$$A_{surface} = \frac{\pi \times D \times L}{2} \tag{14}$$

where L is total length of micro-channels equal to 675 mm = 0.675 m.

From energy balance power supply (P) to the heat sink is equal to the heat absorbed (H) by the micro-channels assuming there are no any heat losses.

Wall temperature (T_w) can bc formulated as

$$T_w = T_{in} + \frac{P}{h_{slurry} A_{surface}} \tag{15}$$

4. Validation of reeult

Validation of result has been done according to Gaikwad and More [15] for Nusselt number (Nu) and flow rate of PCM slurry in the range of 75 ml/min to 300 ml/min at concentration of φ = 0.25 (**Table 5**). Present result at φ = 0.25 compared with Gaikwad and More [15] as shown in **Figure 3**. The results of present work shows similar pattern of variation in Nu with increasing flow rate.

Flow rate (ml/min)	T_{out}		Re		Nu		h_{slurry}		T_w	
	φ=0.25	φ=0.5	φ=0.25	φ=0.5	φ=0.25	φ=0.5	φ=0.25	φ=0.5	φ=0.25	φ=0.5
75	310.744	308.181	2496.58	126.64	12.204	5.474	22,206.27	9729.28	314.52	329.11
100	308.796	306.878	3334.79	169.16	13.942	6.507	26,627.41	11,565.29	312.61	325.13
125	307.654	306.113	4153.51	210.69	15.964	7.418	30,556.78	13,184.47	311.37	322.41
150	306.872	305.590	4992.20	253.23	16.993	8.279	34,291.75	14,714.78	310.46	320.39
175	306.315	305.217	5830.90	295.77	18.526	9.083	37,798.01	16,143.78	309.77	318.85
200	305.907	304.944	6649.61	337.33	20.746	9.825	41,045.81	17,462.58	309.23	317.65
225	305.581	304.727	7488.31	379.85	21.333	10.547	44,218.13	18,745.84	308.78	316.65
250	305.321	304.553	8327.00	422.39	22.456	11.246	47,262.38	19,988.21	308.41	315.80
275	305.113	304.414	9145.71	463.92	22.915	11.884	50,123.64	21,122.17	308.10	315.11
300	304.936	304.295	9984.41	506.47	23.671	12.523	52,959.75	22,257.91	307.83	314.50

Table 5.
Calculated values on the variation of flow rate.

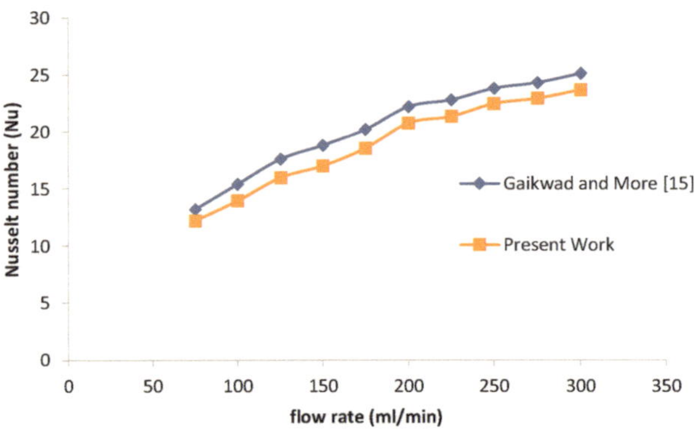

Figure 3.
comparison of Nu vs. flow rate with the results of Gaikwad and More [15].

5. Results and discussion

In present analysis analytical study of square chip of heat sink (25 mm × 25 mm) consisting of semi-circular micro-channels has been carried out using PCM slurry as working fluid with inlet temperature of 303 K. Analytical study is based on the general equations and correlations based on the available literatures. Results obtained for varying mass flow rate from 75 ml/min to 300 ml/min [15]. Effect of flow rate on slurry outlet temperature of heat sink can be seen from **Figure 4** where slurry temperature at the outlet of micro-channel heat sink decreases with mass flow rate, as the specific heat of PCM slurry increases with increasing concentration of PCM particles. Hence at concentration of φ = 0.5 has lower outlet temperature than φ = 0.25 with increasing flow rate and maximum decrement in outlet temperature of slurry is found to be about 0.63% in φ = 0.25 and 0.45% in φ = 0.5. In **Figure 5**. Reynolds number for φ = 0.5 is far below than φ = 0.25. The graph of Reynolds number is almost flat for φ = 0.5 as Reynolds increases marginally with increasing flow rate while for φ = 0.25 the Reynolds number increases linearly with increasing flow rate. This trend signifies that with increasing concentration the viscosity of slurry also increases resulting in decreasing Reynolds number.

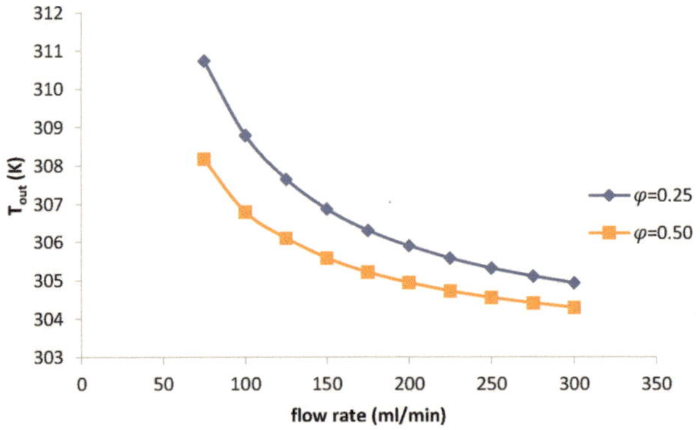

Figure 4.
Effect of flow rate on slurry outlet temperature of the heat sink.

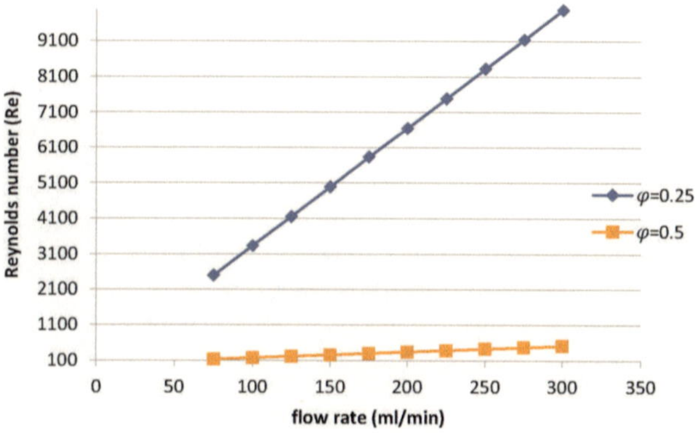

Figure 5
Effect of mass flow rate on Reynolds number.

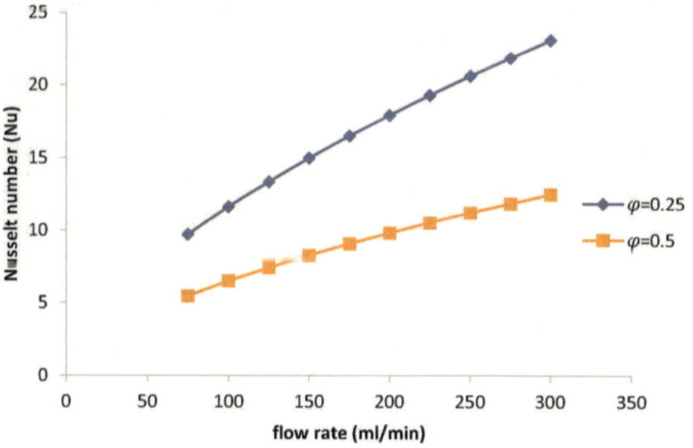

Figure 6.
Effect of mass flow rate on Nusselt number (nu).

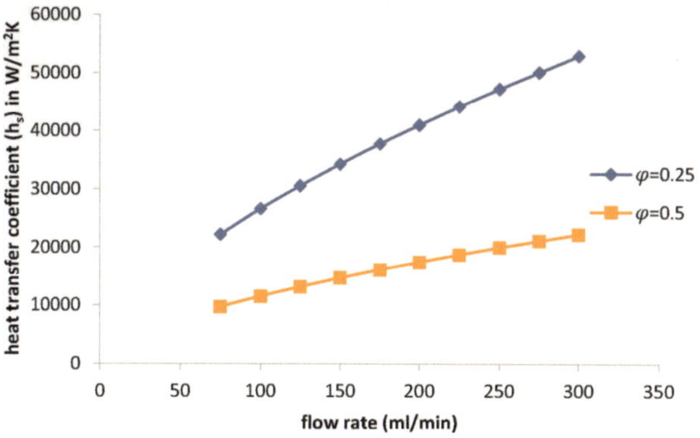

Figure 7.
Effect of mass flow rate on heat transfer coefficient of slurry.

In **Figure 6** Nusselt number (Nu) increases significantly with increase in mass flow rate for both concentrations due to increase in specific heat accordingly. Increase in specific heat of PCM slurry observed because of increase in latent heat of the slurry as

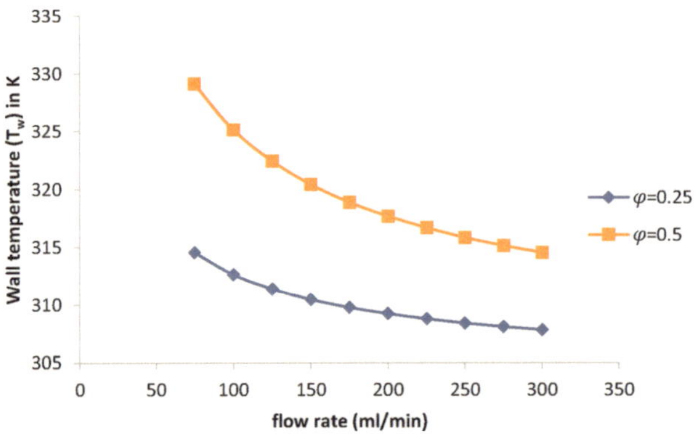

Figure 8.
Effect of mass flow rate on wall temperature of micro-channel heat sink.

it approaches the melting temperature. Higher Nu has been observed for φ = 0.25 while lower value of Nu observed for φ = 0.50 due to high viscous flow at higher concentration of PCM slurry in micro-channels. In **Figure 7** similar trend of graph has been observed for heat transfer coefficient of slurry when flow rate increases as Nu is a function of heat transfer coefficient and maximum increment in heat transfer coefficient of slurry found to be 19.9 for φ = 0.25 and 18.8% for φ = 0.50. In **Figure 8** with increase of the flow rate the wall temperature of the micro-channel heat sink decreases for both concentrations of PCM slurry. Wall temperature decreases significantly for φ = 0.50 than φ = 0.25 due to higher latent heat absorption capacity at higher concentration of PCM particles.

6. Conclusions

In present work analytical study of micro-channel heat sink has been performed at a constant power supply of 95 W using PCM slurry at φ = 0.25 and φ = 0.50 for flow rate in the range of 75 ml/min to 300 ml/min at inlet slurry temperature of 303 K. following conclusions have been drawn.

- Outlet temperature of slurry decreases with increasing flow rate and increasing concentration of PCM particles due to increase in specific heat. Maximum decrement in outlet temperature of slurry observed to be about 0.63% for φ = 0.25 and 0.45% for φ = 0.5.

- Higher viscous flow has been observed for higher concentration of PCM particles results in lower Reynolds number which can be concluded as tending towards laminar flow for higher concentration in micro-channels.

- Nusselt number (Nu) and heat transfer coefficient of slurry increases with flow rate. High heat transfer coefficient and Nu has been observed for φ = 0.25 than φ = 0.50 due to high viscous flow at higher concentration of PCM slurry in micro-channels.

- Wall temperature of micro-channel heat sink which signifies the heat sink temperature decreases with increasing flow rate. Higher percentage decrease of

1.2% in wall temperature has been observed for higher concentration ($\varphi = 0.5$) corresponding to 0.6% decrease in wall temperature for $\varphi = 0.25$ due to higher heat transfer coefficient of highly concentrated slurry.

Author details

Sunny Chandra[1*], Om Prakash[2] and Anupam Prakash[1]

1 Government Engineering College, Banka, India

2 National Institute of Technology Patna, India

*Address all correspondence to: sunnychandra27@gmail.com

References

[1] Brozek T. Micro- and Nanoelectronic s: Emerging Device Challenges and Solutions. Florida, US: CRC Press; 2017. p. 117. ISBN: 9781351831345

[2] "June 2015 Intel CPU microcode update for Windows". [Accessed: November 7, 2020]

[3] Chandra S, Prakash O. Heat transfer in microchannel heat sink: Review. In: International Conference on Recent Advances in Mechanical Engineering (RAME 2016); 14–15 October 2016; DTU. Delhi: RAME; 2016. pp. 571-577

[4] Debich B, Hami AE, Yaich A, Gafsi W, Wallah L, Haddar M. Design optimization of PCM-based finned heat sinks for mechatronic components: A numerical investigation and parametric study. Journal of Energy Storage. 2020; **32**:1-11. DOI: 10.1016/j.est.2020.101960

[5] Chandra S, Prakash O. Impact of channel geometries and flow patterns on Micro-Channel heat sink performance. International Journal of Innovative Technology and Exploring Engineering (IJITEE). 2019;**08**:331-336. DOI: 10.35940/ijitee.K1336.0981119

[6] Hasan MI, Tbena HL. Using of phase change materials to enhance the thermal performance of micro channel heat sink. Engineering Science and Technology, an International Journal. 2018;**21**:517-526. DOI: 10.1016/j.jestch.2018.03.017

[7] Deng X, Wang S, Wang J, Zhang T. Analytical modeling of microchannel heat sinks using microencapsulated phase change material slurry for Chip cooling. In: 10th International Symposium on Heating, Ventilation and Air Conditioning (ISHVAC 2017), Oct. 2017. Vol. 205. Jinan, China: Procedia Engineering; 2017. pp. 2704-2711. DOI: 10.1016/2017.10.194

[8] Xu C, Xu S, Eticha R. D: Experimental investigation of thermal performance for pulsating flow in a microchannel heat sink filled with PCM (paraffin/CNT composite). Energy Conversion and Management. 2021;**236**:1-15. DOI: 10.1016/j.enconman.2021.114071

[9] Chang TC, Lee S, Fuh YY, Peng YC, Zhi-Yu, Lin ZY. PCM based heat sinks of paraffin/nanoplatelet graphite composite for thermal management of IGBT. Applied Thermal Engineering. 2017;**112**: 1129-1136. DOI: 10.1016/j.applther maleng.2016.11.010

[10] Karaipekli A, Bicer A, Sari A, Tyagi VV. Thermal characteristics of expanded perlite/paraffin composite phase change material with enhanced thermal conductivity using carbon nanotubes. Energy Conversion and Management. 2017;**134**:373-381. DOI: 10.1016/j.enconman.2016.12.053

[11] Rostamian F, Etesami N, Haghgoo M. Management of electronic board temperature using heat sink containing pure and microencapsulated phase change materials. International Communications in Heat and Mass Transfer. 2021;**126**:1-9. DOI: 10.1016/j.icheatmasstransfer.2021.105407

[12] Siyabi AI, Khanna S, Mallick T, Sundaram S. Multiple phase change material (PCM) configuration for PCM-based heat sinks-an experimental study. Energies. 2018;**11**:1-14. DOI: 10.3390/en11071629

[13] Kothari R, Mahalkar P, Sahu SK, Khundalwal SI. Experimental Investigations on Thermal Performance of Pcm Based Heat Sink for Passive Cooling of Electronic Components, Proceedings of the ASME 2018 16th

International Conference on Nanochannels, Microchannels, and Minichannels (ICNMM 2018); 10–13 June 2018. Dubrovnik. Croatia: ASME; 2018. pp. 1-9

[14] Thomas J, PVSS S, Krishnan SR, Baby R. Thermal performance evaluation of a phase change material based heat sink: A numerical study. Procedia Technology. 2016;**25**: 1182-1190. DOI: 10.1016/j.protcy.2016.08.237

[15] Gaikwad VP, More SP. PU processor cooling by using microchannel heat sink with PCM as coolant. International Journal of Engineering Research & Technology (IJERT). 2017;**06**:214-219. DOI: 10.17577/IJERTV6IS110119

[16] Kondle S, Alvarado JL, Marsh C. Laminar flow forced convection heat transfer behavior of a phase change material fluid in microchannels: ASME. Journal of Heat Transfer. 2013;**135**: 052801-052811. DOI: 10.1115/1.4023221

[17] Kuravi S, Kota K, Du J, Chow LC. Numerical investigation of flow and heat transfer performance of Nano-encapsulated phase change material slurry in microchannels. Heat Transfer-Transactions of The Asme – Journal of Heat Transfer. 2009;**131**:062901-062909. DOI: 131.10.1115/1.3084123

[18] Memon SA, Sajid MB, Malik MS, Alquaity ABS, Rehman MMU. Investigation of the thermal performance of salt hydrate phase change of nanoparticle slurry flow in a microchannel. Hindawi Journal of Chemistry. 2019;**2019**:1-10. DOI: 10.1155/2019/5271923

[19] Tsuzuki N, Utamura M, Ngo TL. Nusselt number correlations for a microchannel heat exchanger hot water supplier with S-shaped fins. Applied Thermal Engineering - APPL THERM ENG. 2009;**29**:3299-3308. DOI: 10.1016/j.applthermaleng.2009.05.004

Section 5

Phase Change Material for Smart Textile

Chapter 7

The Effect of Laundering on the Physical and Thermal Properties of Phase Change Textile Materials

Ricardo Scheepers

Abstract

Phase change materials (PCMs) have been integrated into clothing and textiles to provide added value in terms of thermoregulation and thus added comfort to the wearer in intensely hot and cold weather conditions. Since clothing is laundered several times during the service life of the textile, the effects of simulated domestic laundering on the physical and thermal properties of a non-woven textile that contains PCMs were investigated. The thermal properties, such as the thermal degradation, melting, evaporation, and crystallisation were measured by the Thermal Gravimetric Analysis (TGA) and the Differential Scanning Calorimetry (DSC) analytical techniques. The physical properties, such as the microcapsule size distribution, and the microcapsule morphologies of the PCM microcapsules, were measured by the Scanning Electron Microscopy (SEM) and Raman analytical techniques. The primary objective of this study is to determine the effect of repeated laundering at different wash temperatures.

Keywords: phase change materials, repeated laundering, thermal degradation, crystallisation, enthalpy, thermograms, thermal properties, physical properties

1. Introduction

Phase change materials (PCMs) are 'latent' heat storage materials. The thermal energy transfer happens when a material changes from solid to liquid or liquid to solid. PCM absorbs and releases heat at an almost constant temperature. Latent heat stores 5 to 14 times more heat per unit volume than sensible storage materials such as bricks or rocks. Latent heat storage is one of the highly effective techniques of storing thermal energy. Unlike the sensible heat storage method, the latent heat storage method requires a significantly higher storage density with a minor temperature difference between storing and releasing heat [1].

As described by Grynaeus et al. [2], PCMs are designed to utilise latent heat absorption associated with a reversible phase change transition, such as solid to liquid transition. The material can be used as an absorber of heat whereby several thermal energies will be absorbed by the PCM before its temperature can increase. The PCM can also be preheated and used as a barrier to cold, as a greater quantity of heat must be removed from the PCM before its temperature can begin to decline.

The development and application of PCM, as a class of thermal energy storage systems, are receiving increasing attention due to their contribution to a more efficient environmentally friendly energy utilisation. The PCM technology was first developed in the 1970s by the National Aeronautics and Space Administration (NASA) of the United States to protect delicate instruments in outer space from large temperature extremes. Today PCMs are one of the most widely used energy storage materials in the fields of solar energy utilisation, energy-conserving in buildings, thermal insulation, and thermal regulation [3].

According to Yazdi & Sheikhzadeth [1], textiles are one of the most used applications of PCMs for thermoregulating principles. During the phase change, the temperature remains unchanged; as a result, the PCM can stabilise the human body temperature. The cooling effect of the PCM depends on the ability to absorb heat during the periods when external heat load or body heat production surpasses heat loss.

In the highly competitive textile marketplace, customers are requiring not only comfort and fashion, but also functionality in both daily wear and sporting apparel. Phase change materials have been incorporated into clothing to provide added functionality and value in terms of thermoregulation and thus increased comfort to the wearer, in notably extreme weather conditions and sporting activities. Since clothing is laundered many times, it was deemed important to study the effect of repeated laundering on the thermoregulating properties of fabrics containing PCMs.

In this study, the effect of simulated domestic laundering on the physical and thermal characteristics of a non-woven textile that contains PCM is investigated since this is important in terms of the functional lifecycle and durability of PCMs and can contribute to improving the physical and thermal performance of PCMs and their design, as well as the laundering conditions [4, 5].

2. Experimental

2.1 Sample preparation (according to EN ISO 6330:2000 standard)

- Lay the fabric to relax for at least 4 hours in a standard atmosphere (temperature of 20°C ± 2°C and relative humidity of 65% ±2%).

- Cut out the samples using a template and mark accordingly.

- Overlock all edges to avoid fraying of the samples at the edges.

2.2 Washing procedure (according to EN ISO 6330:2000 standard)

- Wash load: Ensure no more than half of the load are consist of the test samples, the rest of the load being made up of makeweights.

- Detergent: Use 5 grams of European Colourfastness Establishment (ECE) detergent and 2 grams Sodium Perborate. ECE detergents do not contain any optical bright agents (OBA).

- Wash Program:

a. Select the correct program for the type of wash required. Set temperature and water levels. Place makeweights and then the test samples in the washing machine. Dissolve the detergent with a small quantity of warm water in a beaker.

b. Start the machine. When the water is above the base level in the sight tube, add the dissolved detergent into the machine by lifting the lid on top of the machine.

c. On completion of the program remove the wash load.

2.3 Drying method (according to EN ISO 6330:2000 standard)

Flat dry – place the samples on a drying rack without stretching the samples. Wait at least 24 hours until completely dry before assessing samples.

3. Testing equipment and testing procedures

3.1 TGA analytical instrument (according to ISO 11358 standard)

TGA test procedure was followed where 10 to 15 mg of samples were heated from 30 to 500°C at a heating rate of 20°C/min under nitrogen atmosphere.

3.2 DSC analytical instrument (according to ISO 11357-1:2016 standard)

The DSC instrument was used to evaluate the latent heat of the PCM samples. The samples were cut into small pieces with a sharp blade after which they were spread evenly on the surface of the pans. Some 10 to 15 mg of samples were added to the pan and weighed on the balance which measures up to the accuracy of 0.0001 mg.

3.3 DSC testing procedure

- Turn on the machine and allow it to warm up for about an hour.
- Ensure the compressed nitrogen tank and liquid nitrogen tank are both full and the valve connection is open. Set the compressed nitrogen pressure flow.
- Prepare empty pans and place the pans on the two circular sensors within the furnace.
- Activate the machine software and input all necessary information in the program.
- Input a temperature 10°C higher than the highest temperature set in the temperature program. This is a protective setting, which stops the machine from surpassing a set temperature in case of machine malfunction.
- To start the program, the current furnace temperature must be 5°C of the initial temperature.
- Start and the measurement will begin, after baseline scan has run, remove the empty baseline pan, and replace it with the pan containing the samples.

- To ensure even heat flow and accurate DSC readings, a thin layer of sample pieces is placed so the entire bottom of the pan is covered.
- After the measurement has finished, close the program, turn off the compressed nitrogen tank and turn off the machine.

3.4 Scanning electron microscopy (SEM)

3.4.1 SEM sample preparation (according to ISO 22493:2014 standard)

- A small amount of the fabric was mounted on a 12 mm aluminium SEM stub using carbon-coated glue.

3.4.2 SEM analysis procedure

SEM settings: Backscatter detector, 20 kV accelerating voltage, working distance of 10 mm. Magnification between 1000 and 4000x.

- Insert the sample into the chamber of the SEM and observe in low vacuum mode.
- SEM is equipped with lenses to compress the spot and direct the focused electrons on the sample.
- The image of the sample is generated point by point dependent on the movement of the scan coils, which causes the electron beam to move to positions in a form of straight lines until a rectangular construction is formed on the surface of the sample.
- If necessary, change to a higher magnified image, and the scan coils will then make the beam deflect a smaller area.
- The electron detector detects the emitted electrons (signals) from the scanned sample.
- The signals are then displayed on the viewing screen, the brightness and the intensity can be controlled until a clear image is obtained.

3.5 Raman microscopy and spectroscopy

3.5.1 Raman sample preparation

Raman analysis needs no sample preparation. Images of the fibres were taken with the SEM after which the sample was moved from under the SEM pole piece to the Raman objective in the SEM chamber and a Raman map was acquired of the same area.

3.5.2 Raman analysis procedure

Raman Settings: Laser wavelength 532 nm, the laser power was 2 mW, map areas were 50μmx50μm with data points every 0.5 μm, integration time (time spent on each point) 0.2 seconds.

- Insert the sample, and make sure it is flat enough to avoid scratching the lenses.
- Repeat the position and focus procedure until the sample is focused on the desired magnification.
- Adjust the instrument parameters (exposure time, laser filter, spectrometer off-set, etc.) until the desired spectrum is shown on the display and capture Raman Image and Spectrum.
- Press stop and turn off the lamp.

4. Analysis

Comparative analysis was applied to the DSC, TGA, SEM and Raman analytical results.

5. Results and discussions

The main objective of this study is to determine the effect of repeated laundering at different wash temperatures, on the thermal and physical related properties of the PCM as measured by TGA, DSC, SEM and Raman.

5.1 Thermal gravimetric analysis (TGA)

5.1.1 TGA results of multiple washed samples at different temperatures

Thermal gravimetric analysis (TGA) is generally used to investigate chemical reactions in which weight changes occur. Most TGA curves illustrate mass losses, for example in **Figures 1-3** the TGA thermograms show 3 different degradation regions that indicate a combination of paraffin wax that forms the core of the microcapsule, an acrylic resin as the polymeric shell of the microcapsule and polymeric fibres which can be identified as Polyester (PES), Polyamide (PA) and Polyethylene Terephthalate (PET) present in the PCM sample.

5.1.2 Mass loss percentage (%) comparison of different wash temperatures after 10 washes

In **Figure 4** the thermogram of the first segment was identified as the paraffin wax, where it is observed for the sample washed at 30°C there is a mass loss of 15.8% at a heating temperature of around 208°C, for the sample washed at 40°C there is a mass loss of 18.8% at a heating temperature of approximately 205°C and for the sample washed at 60°C a mass loss of 19.1% was observed at a heating temperature of nearly 200°C. The thermogram indicate that the mass loss at wash temperature 60°C is slightly higher with a lower heating temperature compared to the wash temperature of 30°C and 40°C where the mass loss is lower at a higher heating temperature.

The thermogram of the second degradation segment, was identified as the acrylic resin of the microcapsule, which exhibit a mass loss of 24.5% of the sample at wash temperature 30°C, a mass loss of 31.8% of the sample at wash temperature 40°C and

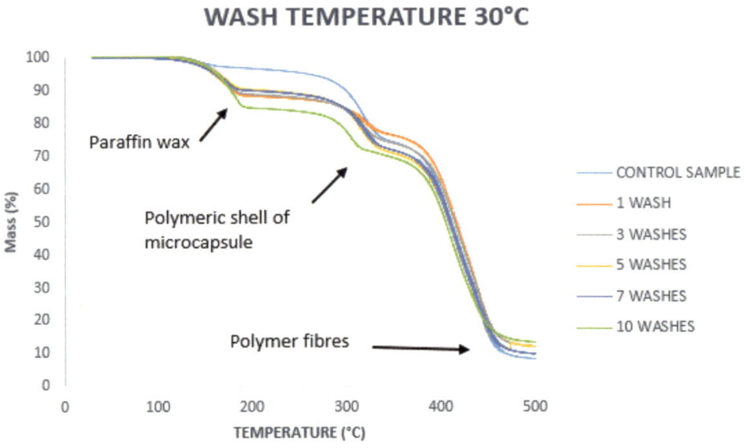

Figure 1.
TGA thermograms showing the effect of repeated laundering on mass loss (number of washes at 30°C).

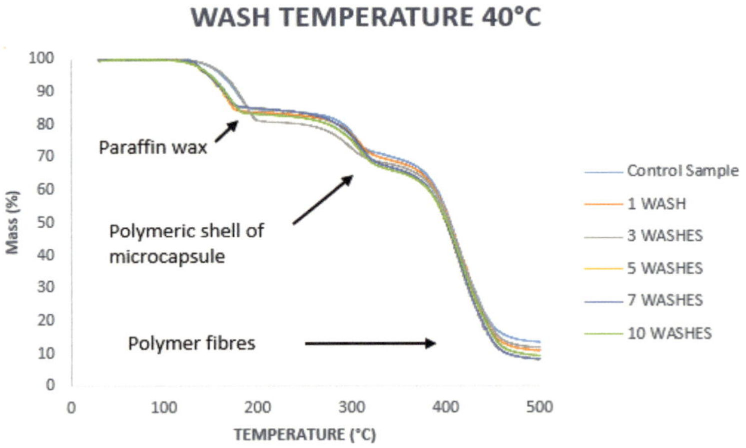

Figure 2.
TGA thermograms showing the effect of repeated laundering on mass loss (number of washes at 40°C).

a mass loss of 35.1% of the sample at wash temperature 60°C with heating temperatures ranging from 300 to 400°C. This proves that a higher mass loss occurred for the sample at wash temperature 60°C compared to the samples of the lower wash temperatures, this is due to the PCM microcapsules breaking during the laundering process.

The last segment which indicates the thermal degradation of the polymeric fibres (PES, PA, and PET) show a mass loss of around 87.5% ±2% for the samples washed at temperatures 30°C, 40°C and 60°C at a heating temperature up to 500°C.

The difference in onset degradation is caused by the different wash temperatures where loss of mPCM content occurred during the laundering process. The thermogram clearly indicate that there is a higher mass loss at wash temperature 60°C, than at wash temperature 30°C and 40°C, respectively. The difference in average mass loss percentage between the sample at wash temperature 60°C compared to the sample at wash temperature 30°C is about 4.6%.

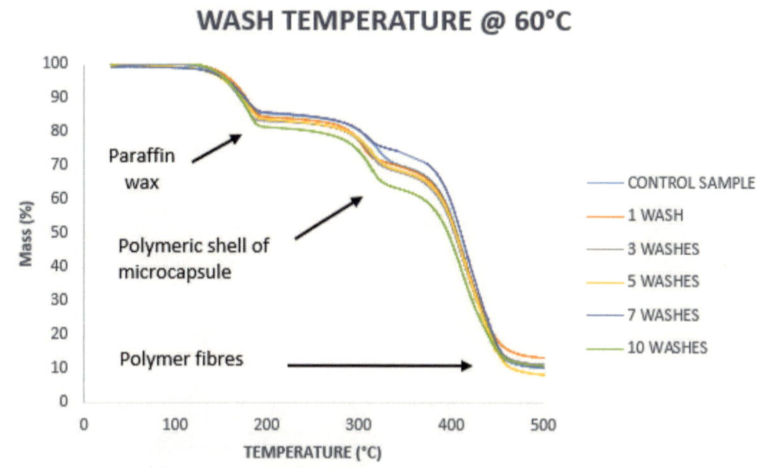

Figure 3.
TGA thermograms showing the effect of repeated laundering on mass loss (number of washes at 60°C).

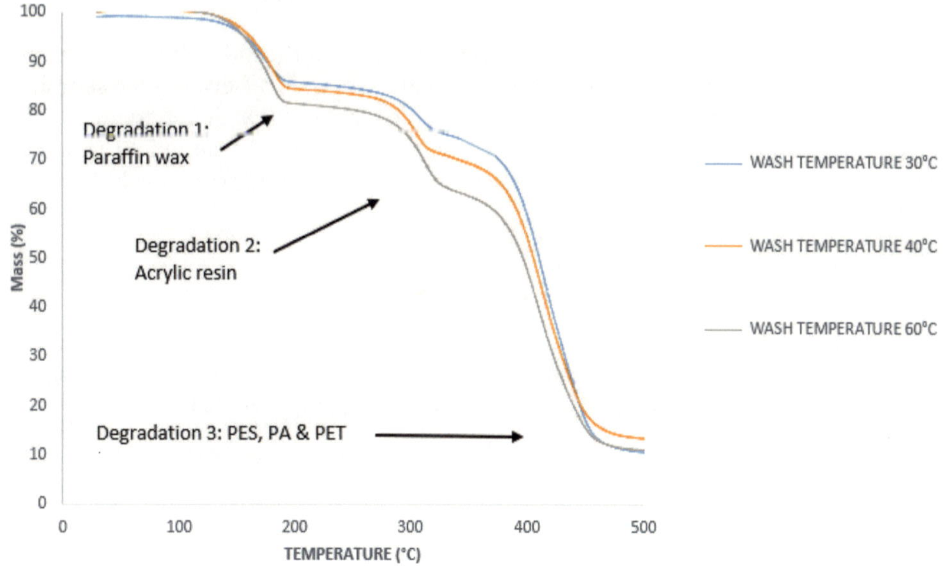

Figure 4.
TGA thermograms illustrating the effect of wash temperatures on mass loss after 10 washes.

5.1.3 Mass loss percentage (%) of unwashed sample vs. multiple washed sample at 60°C

Figure 5 shows a transition at around heating temperature of 150°C for the sample at wash temperature 60°C, which is not present for the unwashed sample. The onset degradation for the sample washed at 60°C, differs with a mass loss percentage of 26.12% compared to the unwashed sample. The difference in mass loss is substantial and is due to the PCM capsules being broken at the wash temperature of 60°C and the associated loss of thermoregulating content, whereas with the unwashed sample, the thermoregulating content was unchanged.

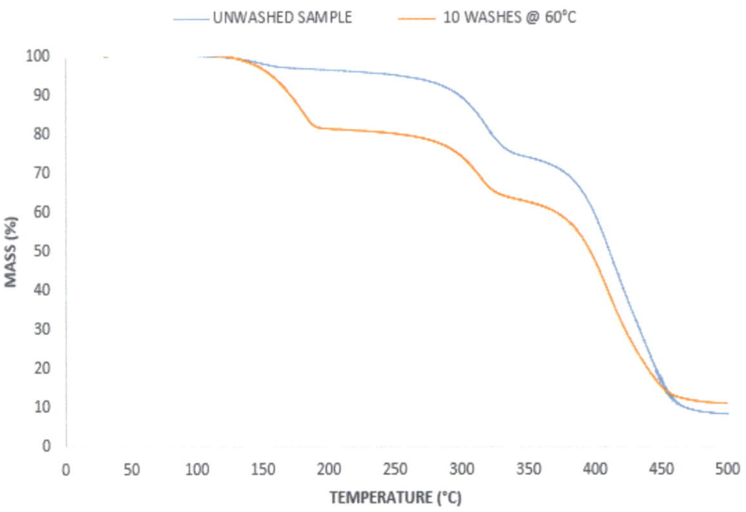

Figure 5.
TGA thermogram demonstrating mass loss for the unwashed sample and a multiple washed sample at 60°C.

The mass loss for each degradation phase, indicate that the mass loss is higher for the sample washed at 60°C. The difference in average mass loss between the sample washed at 60°C compared to the unwashed sample is about 9.4%. It is evident that both samples burn out around the same heating temperature of around 500°C, where the ash content can be calculated as a repeatability average of approximately 12.5%.

5.2 Differential scanning calorimeter (DSC)

The DSC analysis is used to assess the melting point, crystallisation, and heat of fusion to determine the change in thermal properties of the PCM due to the different wash temperatures. The DSC analysis results are presented in the form of thermograms as illustrated in **Figures 6-8**. The primary focus is to evaluate the effect of

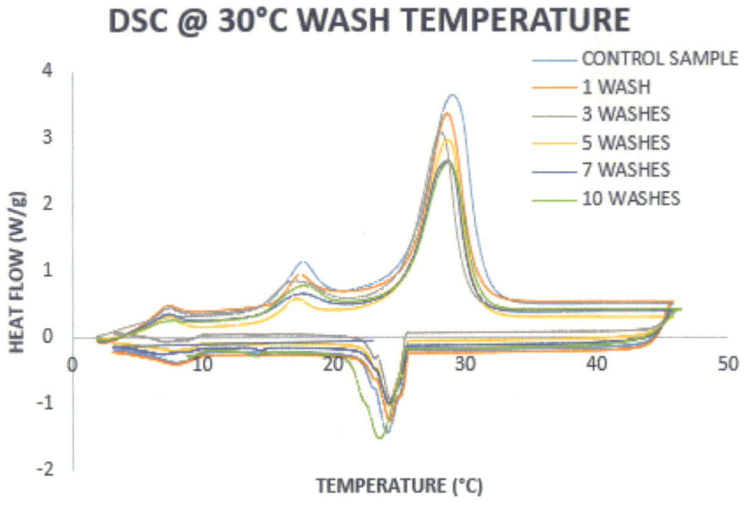

Figure 6.
DSC analysis of washed samples (wash temperature 30°C).

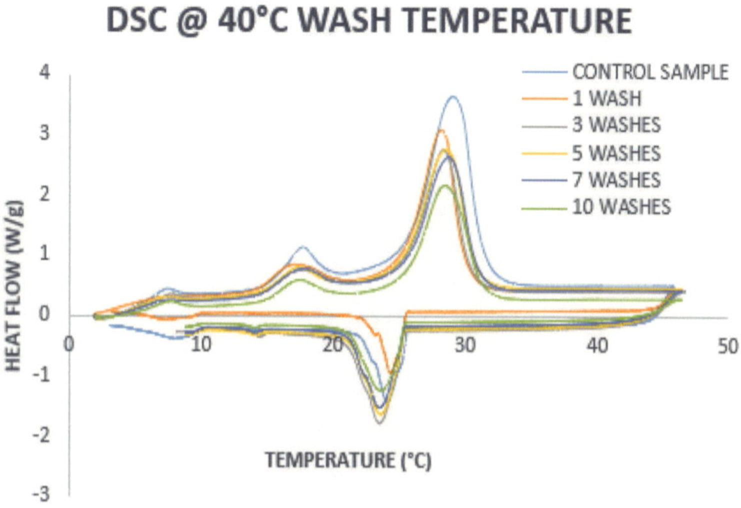

Figure 7.
DSC analysis of washed samples (wash temperature 40°C).

the different wash temperatures in terms of the heating and cooling peaks of wash temperature (30°C, 40°C and 60°C), also of the samples at 10 washes to observe the worst-case scenario.

Figure 9 shows 3 distinct melting peaks for each of the samples at wash temperatures (30°C, 40°C and 60°C), these melting peaks are probably related to the molecular structure of the hydrocarbons in the paraffin wax PCM (tetradecane, hexadecane, and octadecane). The main interest is the enthalpy values of the third melting peaks where the enthalpy of the octadecane content changed, indicating phase transition of the PCM from solid to liquid state at a heat temperature of 28°C ± 2°C for all three of the samples at different wash temperatures.

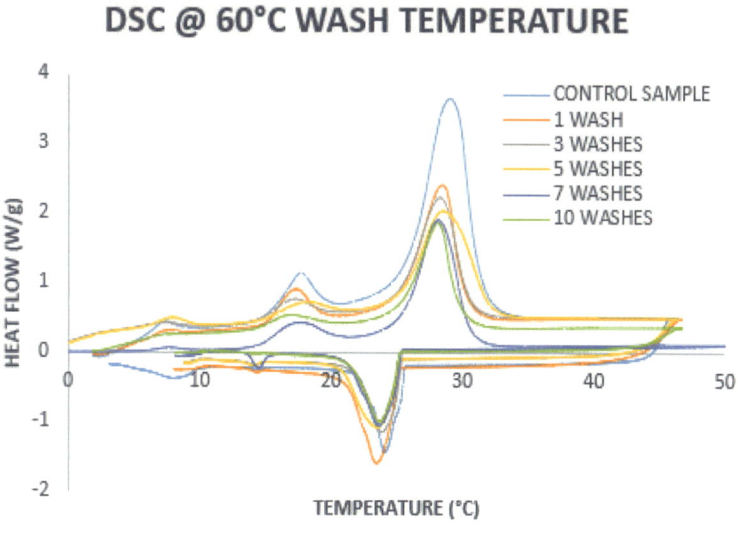

Figure 8.
DSC analysis of washed samples (wash temperature 60°C).

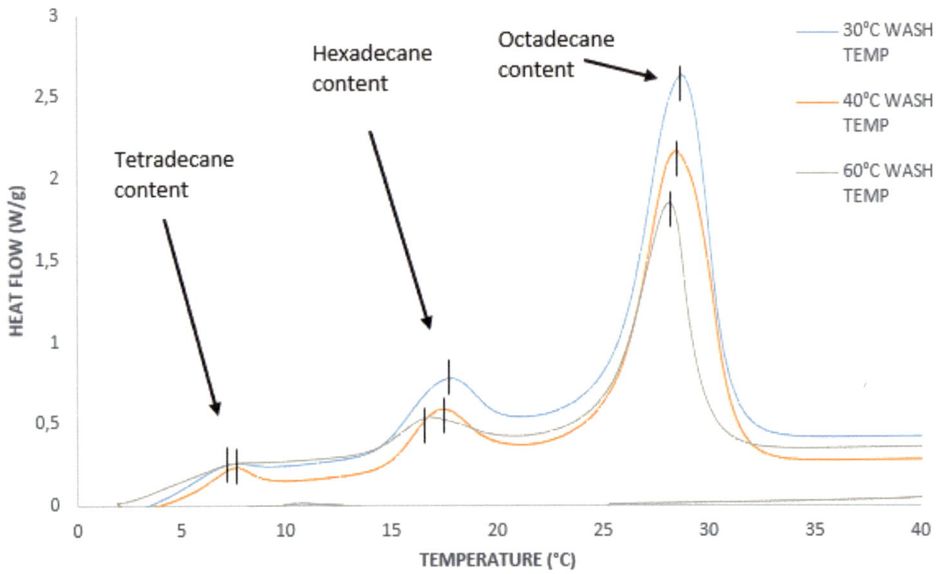

Figure 9.
Endothermic results showing the effect of different wash temperatures on the heat flow for samples washed 10 times.

It is apparent that the heat capacity of 1.8 W/g for the sample washed at 60°C is lower than the sample washed at 30°C, with a heat flow of 2.62 W/g, and for the sample washed at 40°C, with a heat flow of 2.16 W/g. It can be established that the heat loss is greater at wash temperature 60°C, this is due to the loss of thermoregulating content, broken microcapsules, and loss of fibres.

5.2.1 Exothermic results of multiple washed samples at different temperatures

The crystallisation temperature is defined as the lowest point of the cooling peak. The enthalpy of crystallisation is being determined by the area under the curve. **Figure 10** shows details of crystallisation occurring where the cooling peaks of samples washed at different wash temperatures coincide at heating temperatures of 9°C ± 2°C at wash temperature 30°C, 14°C ± 2°C at wash temperature 40°C and 23°C ± 2°C at wash temperature 60°C.

The primary focus is on the third cooling peaks where the heat flow was at its peak. It can be observed that the sample at wash temperature 60°C has significantly less degree of crystallinity with a heat flow of −0.99 W/g compared to the samples at wash temperature 30°C, where the heat flow is −1.48 W/g, and at wash temperature 40°C, where the heat flow is −1.26 W/g. These results could be explained by leakage or evaporation of paraffin after the rupture of the PCM microcapsules during the laundering process at wash temperature 60°C.

5.2.2 Effect of 10 washes on thermal properties

The effect of multiple laundering (10 washes at 60°C) on the thermal properties are shown in **Figure 11**. The multiple washing exhibits lower treatment heat capacity and cooling capacity than the unwashed sample. There is a difference of

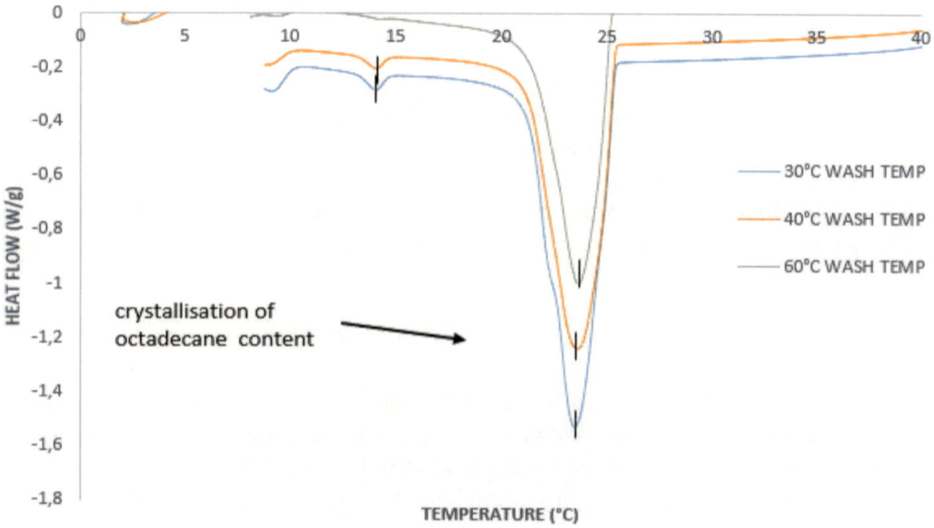

Figure 10.
Exothermic results showing the effect of different wash temperatures on the heat flow for samples washed 10 times.

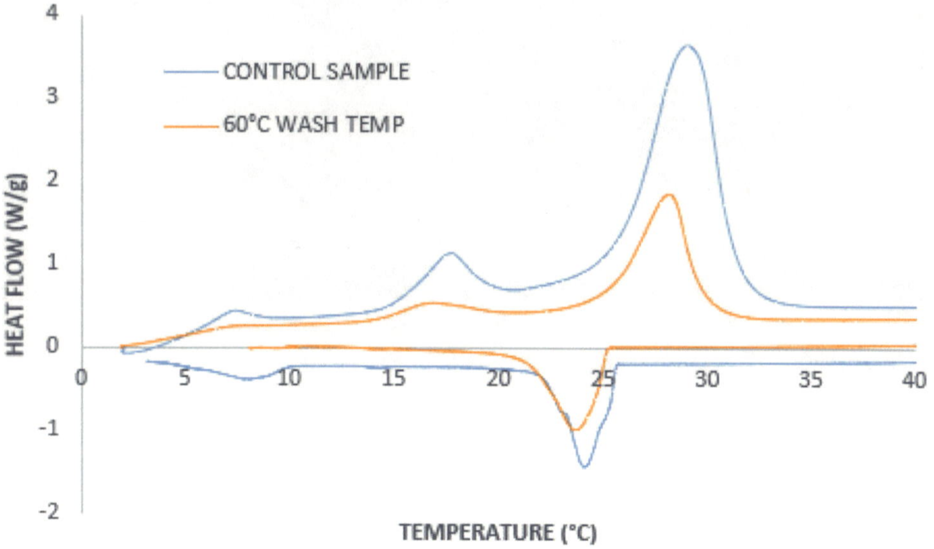

Figure 11.
Effect of 10 washes at wash temperature 60°C compared to the unwashed sample.

approximately 50% latent heat capacity between the unwashed sample and the sample washed at 60°C, this is due to the loss of PCM content after 10 wash cycles at a higher wash temperature.

5.3 SEM analysis

The SEM analysis will discuss the effect of multiple washes at different temperatures on the morphology (shape and size of PCM microcapsule) that was examined by the means of SEM images at different magnifications.

5.3.1 Unwashed PCM: different areas at 500X magnification (100 μm)

SEM images at magnification 500X was captured at 3 different areas on the sample to illustrate distribution of the PCM microcapsules within the sample. In **Figure 12** image A, B and C shows visible bulges on the fibres and in an uneven webbing that links the intercedes of micron sized PCMs. The bulges indicates that the PCM content is wrapped around the fibre and the micron sized PCM observed in the uneven webbing indicates the PCM content was incorporated through a pad-dry-cure process method.

5.3.2 Unwashed PCM: different magnifications (100X, 500X, and 3000X)

In **Figure 13**, image A at magnification 100X shows the web linking oriented fibres of the non-woven PCM sample. Image B at magnification 500X there are 2 different form of PCMs observed, the PCM microcapsules in the web and the mPCM wrapped around the fibre. Image C at magnification 3000X clearly illustrates the PCM capsules within web.

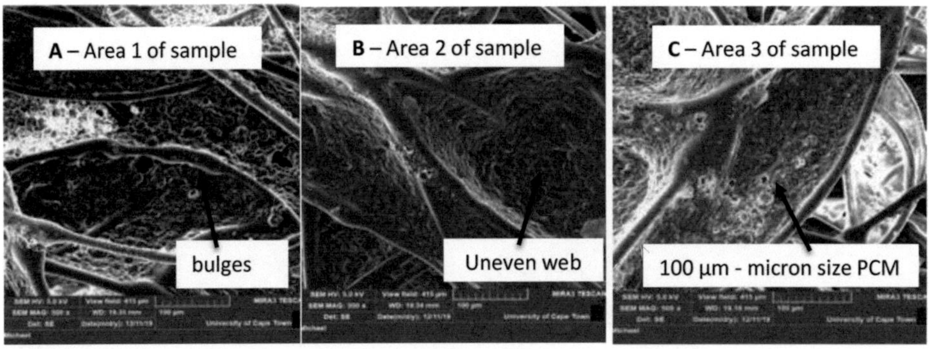

Figure 12.
SEM images of unwashed sample at different areas.

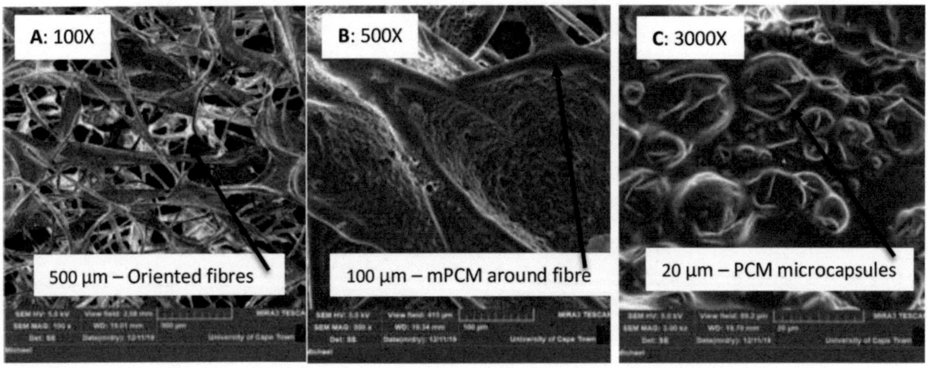

Figure 13.
SEM images of unwashed PCM at different magnifications.

5.3.3 SEM images: Multiple laundered (10 washes) samples at different wash temperatures

5.3.3.1 Sample at wash temperature 30°C

Figure 14 shows that image A at 500X magnification there is fibre interlacing and holes within the web, whereas image B at magnification 3000X clearly shows the broken PCM capsules, which indicate PCM content has evaporated during laundering as indicated by the TGA and DSC results.

5.3.3.2 Sample at wash temperature 40°C

In **Figure 15** the sample at wash temperature 40°C where image A at 500X magnification illustrates brittle fibres in the webbing, and Image B at magnification 3000X

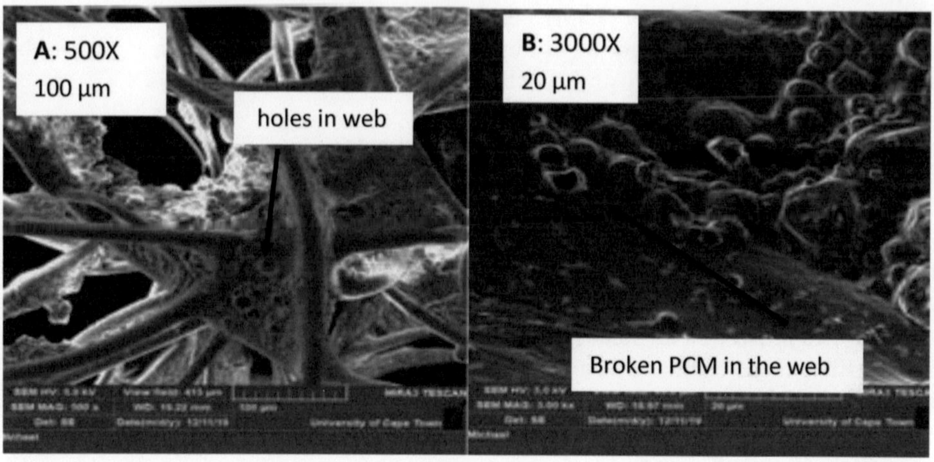

Figure 14.
SEM images: mPCM of sample at wash temperature 30°C.

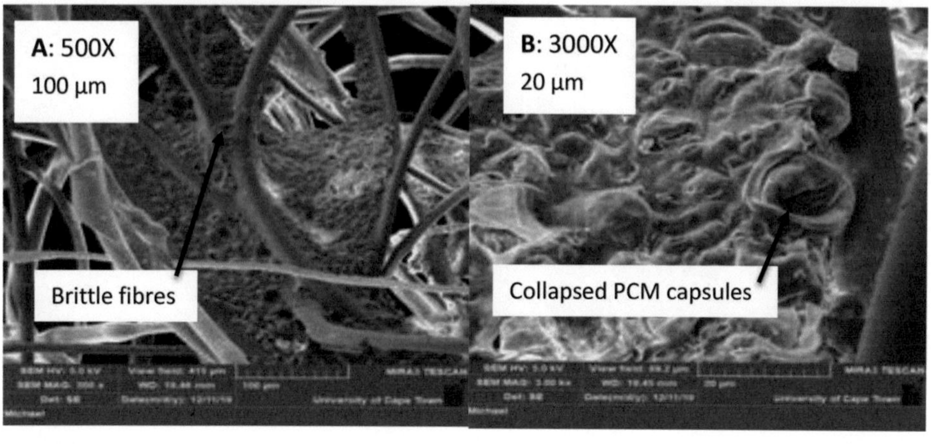

Figure 15.
SEM images: mPCM of sample at wash temperature 40°C.

evidently illustrates collapsed PCM capsules in the web, which is due to multiple laundering at higher wash temperatures as quantified in TGA and DSC analysis.

5.3.3.3 Sample at wash temperature 60°C

In **Figure 16** it can be observed that image A at magnification 1620X the mPCM around the fibre are brittle and fragmented and image B at magnification 3280X it is evident that the PCM capsule around the fibre is damaged, therefore the conclusion can be made that multiple laundering at higher wash temperatures are damaging to the PCM microcapsule, which causes the loss of thermoregulating content as identified in the TGA and DSC test results.

5.4 Raman analysis

The Raman analysis will investigate the effect of multiple laundering at different temperatures on the polymer changes in the crystalline phase through the sample, the crystallinity distribution, and the orientation of the polymers, that was examined by the means of Raman imaging and spectrum.

5.4.1 Raman imaging and Spectrum of unwashed sample

In **Figure 17** the scanning position of image A illustrates the PCM in the web and on the fibre, and image B shows the dept. profiling of the Raman image which displays 4 different wavelengths in the spectrum represented by colours: red, dark blue, green, and light blue.

Figure 18 illustrates the peaks of the Raman spectrum, these shifting peaks can be identified as the chemical bond lengths of the n-alkanes paraffins, namely tetradecane, hexadecane and octadecane, which form the core of the microencapsulated PCM, and the fourth shifting peak observed could be the polymeric shell microcapsule which is an acrylic resin.

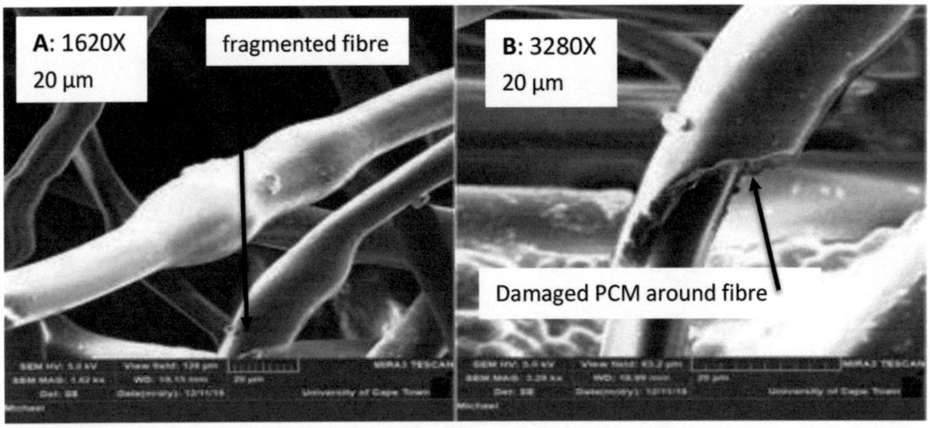

Figure 16.
Damaged mPCM around fibre of sample at wash temperature 60°C.

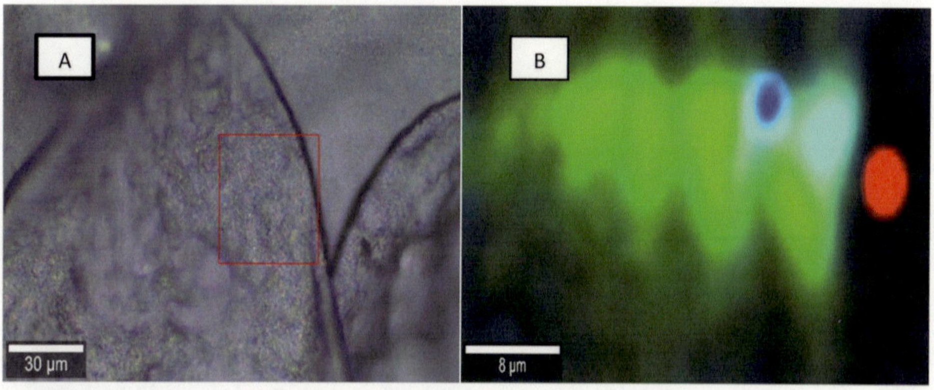

Figure 17.
Raman scan position of unwashed sample (left: Image a); Raman scan depth of unwashed sample (right: Image B).

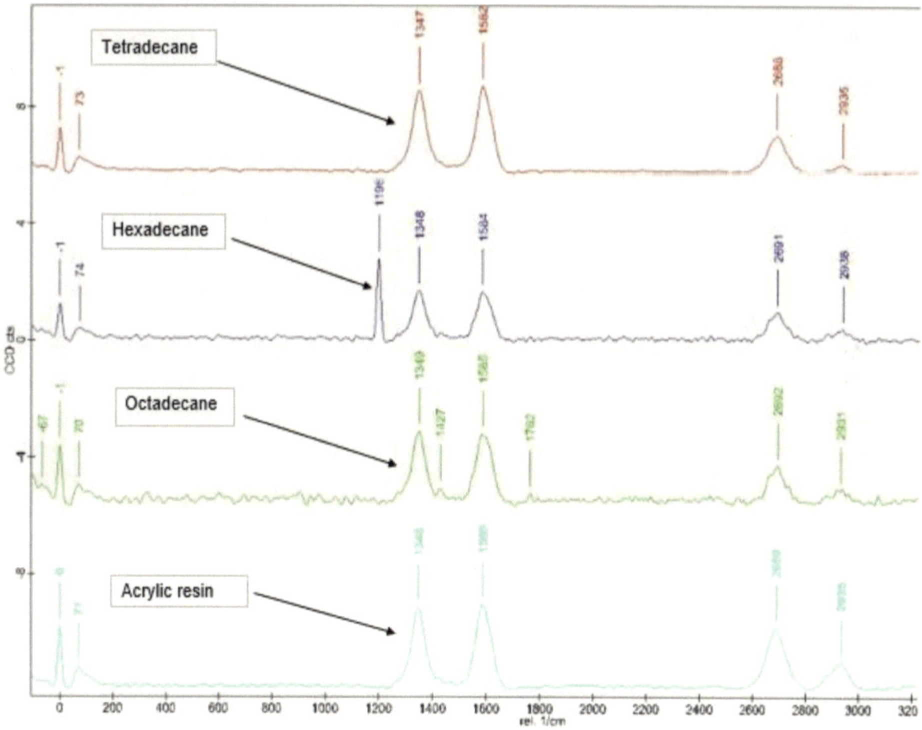

Figure 18.
Raman spectrum of unwashed sample.

The width of the shifting peaks on the graph illustrates a high degree of crystallinity, which is important, because most physical and mechanical properties of PCMs are affected by the degree of crystallinity. The crystallinity distribution in the sample can be found from the band shift area and the orientation of the polymers in the spectrum can be seen.

5.4.2 Raman analysis of multiple washed samples at different temperatures

5.4.2.1 Raman imaging and Spectrum of multiple washed sample at 30°C

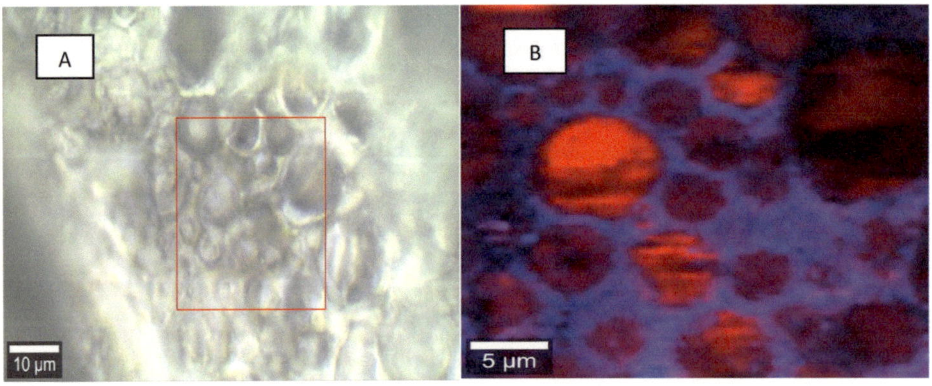

Figure 19.
Raman scan position of washed sample at 30°C (left: Image a); Raman scan depth of washed sample at 30°C (right: Image B).

5.4.2.2 Raman imaging and Spectrum of multiple washed sample at 40°C

In **Figures 19** and **20** the scanning position of images A illustrates the PCM microcapsules in the web and therefore the depth profiling of the Raman in images B only show 2 different wavelengths which are recognised as the n-octadecane paraffin and the acrylic resin in the spectrum represented by colours: blue and red.

Figure 21 represents the Raman spectrum of the washed sample at 30°C and **Figure 22** represents the Raman spectrum of the washed sample at 40°C. In both these spectrums it can be observed that there is a change in band positioning, the width and shape of the peaks have transformed, which can indicate polymer changes in the crystalline phase, as indicated in the DSC analysis. The change in band shift is

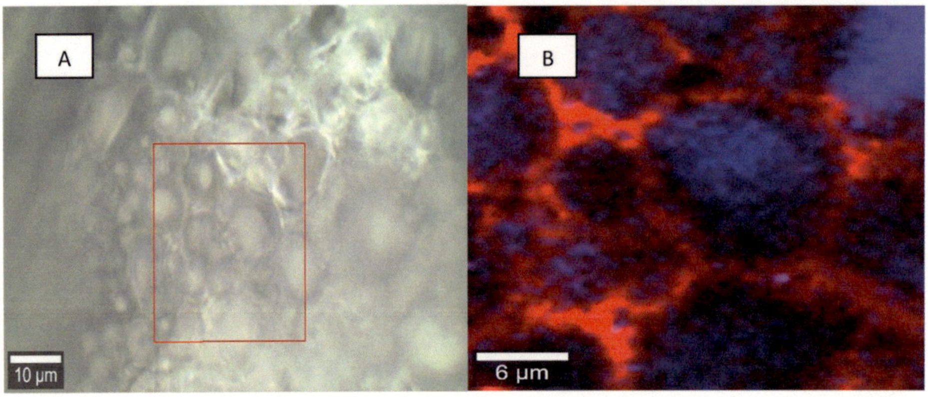

Figure 20.
Raman scan position of washed sample at 40°C (left: Image a); Raman scan depth of washed sample at 40°C (right: Image B).

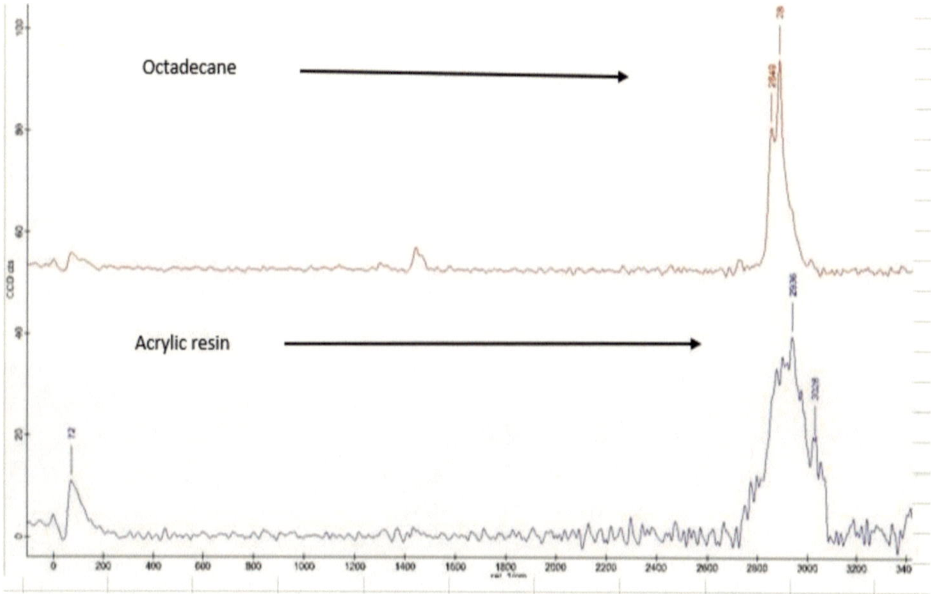

Figure 21.
Raman spectrum of washed sample at 30°C.

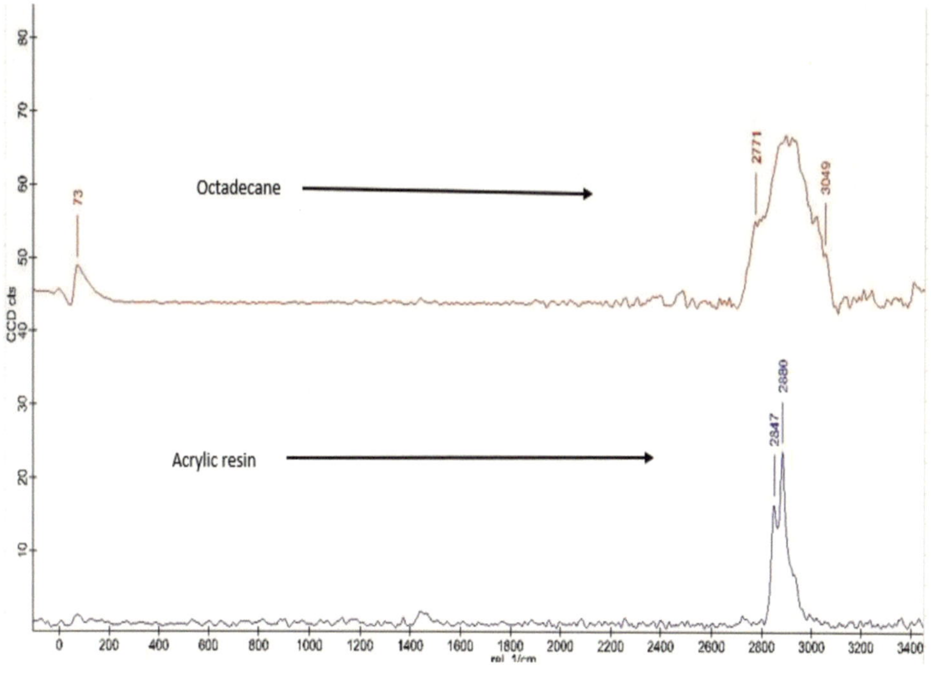

Figure 22.
Raman spectrum of washed sample at 40°C.

also owing to the temperature and stress applied from to the laundering effect, the more orientated the washed sample, the more shift is seen.

5.4.2.3 Raman imaging and Spectrum of multiple washed sample at 60°C

In **Figure 23** the scanning position, image A shows the PCM in the web and on the fibre, however the dept. profiling of the Raman image B only exhibits 3 wavelengths in the spectrum identified as hexadecane, octadecane and the acrylic resin, represented by the colours: red, blue, and green. It can be understood that due to multiple laundering, and rupture of the PCM microcapsules, tetradecane content has evaporated at wash temperature 60°C.

Figure 24 illustrates the Raman spectrum of the washed sample at 60°C, the Raman shift, intensity, peak width, and shape is different to the unwashed sample

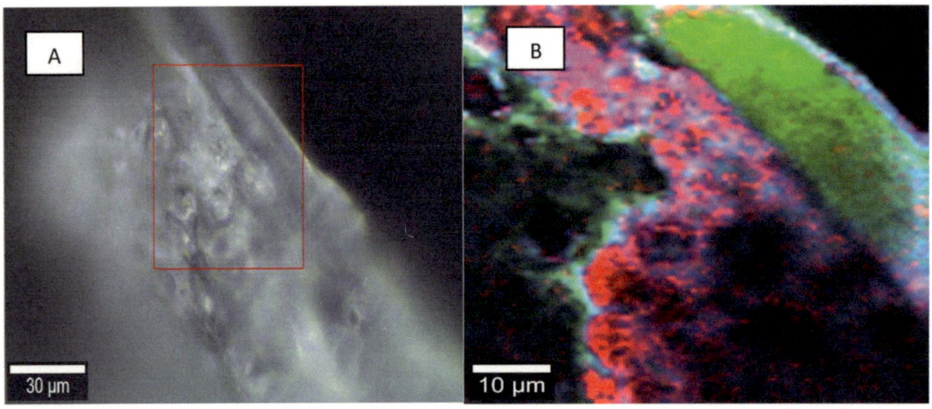

Figure 23.
Raman scan position of washed sample at 60°C (left: Image a); Raman scan depth of washed sample at 60°C (right: Image B).

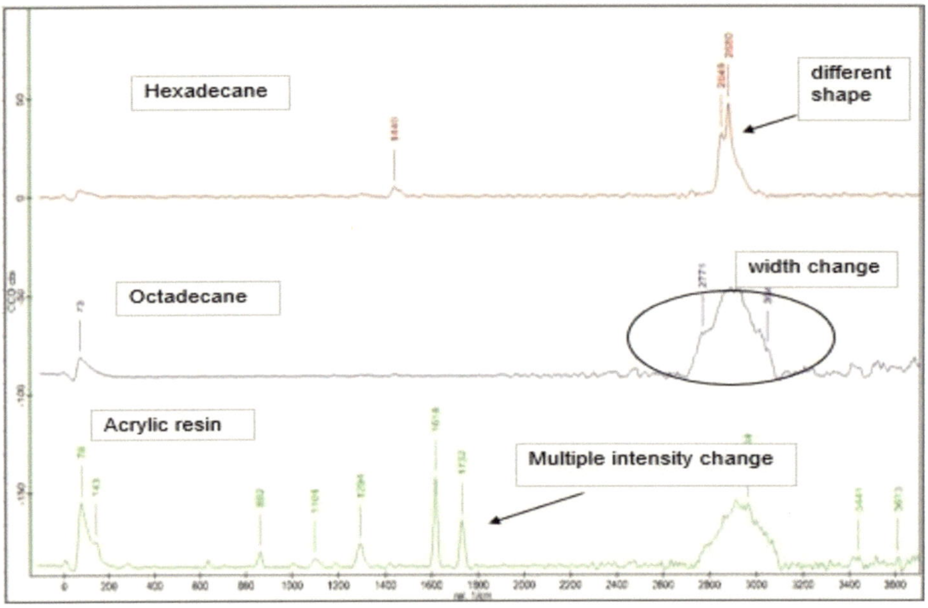

Figure 24.
Raman spectrum of washed sample at 60°C.

and the washed samples at 30°C and 40°C, the chemical composition, crystal size, molecular chain length and morphology vary for different areas of the samples.

It is evident that the washed sample at 60°C show a Raman peak with a wider width and thus indicates a lower degree of crystallinity, which possibly means that the latent heat effectiveness of the PCM has reduced. It is also apparent that the intensity and polarisation of the Raman peaks have changed compared to the unwashed sample, which means the crystallinity and orientation of the PCM has changed due to loss of thermoregulating content, and therefore validate that crystallinity is a structural arrangement in the material and can be calculated from the DSC analysis.

6. Conclusions

The main objective of this study was to investigate the effect of multiple laundering at different wash temperatures, on the thermal and physical related properties, as measured by TGA, DSC, SEM and Raman, of a non-woven textile fabric containing PCM.

Thermal degradation as determined by TGA thermograms was found to be higher at a wash temperature of 60°C relative to that of a wash temperature at 30°C, the mass loss being 4.6% more at 60°C than at 30°C. This was asserted to physical changes such as size and shape of the PCM microcapsules as evident in SEM images. The SEM images showed broken and collapsed PCM microcapsules in the web and around the fibre, indicating evaporation of the thermoregulating content of the PCM.

It was found that the heat capacity of the sample washed at 60°C was much lower than that of the unwashed sample, the difference being approximately 50% in terms of the latent heat capacity as determined by DSC thermograms. This implies that the latent heat effectiveness of the PCM has decreased significantly. The physical properties of the PCM such as the intermolecular interaction, the intensity and width of peaks, polarisation, degree of crystallinity and the orientation changed as illustrated in the Raman analysis.

It can be concluded that the multiple laundering of textiles which contain PCM causes damage to the PCM capsules, causing the thermoregulating effectiveness to be reduced as confirmed by the results from the analytical techniques. In addition, the results also indicate that reducing the wash temperature and improving the design and formation method of the PCM microcapsules can extend the usable life and value of a textile containing PCMs.

Author details

Ricardo Scheepers
Nelson Mandela University, Port Elizabeth, Eastern Cape, South Africa

*Address all correspondence to: rscheeprs0509@gmail.com

References

[1] Yazdi MM, Sheikhzadeth M. Personal cooling garments: A review. The Journal of Textile Institute. 2014;**105**(12):1231-1250

[2] Grynaeus P, Russel D, O'Regan T, Johnson SG, Dietel DS. Thermal control nonwoven material. European Patent 1 587 977 B1. 2013

[3] Yu S, Wang X, Wu D. Microencapsulation of n-octadecane phase change material with calcium carbonate Shell for enhancement of thermal conductivity and serving durability. Synthesis, microstructure and performance evaluation. Applied Energy. 2014;**114**:632-643

[4] Abuzaid AI, Reichard G. An Assessment of Utilizing Phase Change Materials (PCM) Towards Energy Performance in Building Enclosures. USA: Penn State, University Park; 2016

[5] Ahrari M, Khajavi R, Dolatabadi MK, Toliyat T, Rashidi A. A review on application of phase change materials in textile finishing. International Journal of Materials and Metallurgical Engineering. 2017;**11**(5):401

Section 6

Other Applications

Chapter 8

Germanium Telluride: A Chalcogenide Phase Change Material with Many Possibilities

Turja Nandy, Farhana Anwar and Ronald A. Coutu Jr.

Abstract

Germanium telluride (GeTe) is a chalcogenide phase change material which is nonvolatile and changes its phase from amorphous state to a highly conductive crystalline state at approximately 180–230°C temperature, dropping the material's resistivity by six orders of magnitude. These temperature-induced states lead to different physical and chemical properties, making it a suitable candidate for optical storage, reconfigurable circuit, high-speed switching, terahertz (THz), and satellite applications. Besides, GeTe-based devices offer complementary metal oxide-semiconductor (CMOS) compatibility and simplified, low-cost fabrication processes. In this chapter, three applications of GeTe will be discussed. They are as follows: (1) how GeTe can be utilized as DC and RF switching material with their high OFF/ON resistivity ratio, (2) how GeTe can contribute to current THz technology as split-ring resonators and modulators, and (3) effect of threshold voltage on GeTe for reconfigurable circuits.

Keywords: germanium telluride, phase change material, switching, reconfigurable circuit, terahertz technology

1. Introduction

Switching plays the most significant role in circuit technology. Phase change material (PCM) such as germanium telluride (GeTe)-based switching can provide reliable switching performance with fewer mechanical components and higher device lifetime. When an external stimulus is provided to GeTe, it can behave like both conductors and insulators which are dependent on that external stimulus. This phenomenon can be used as DC switching mechanism [1]. It is crucial to study the properties of both amorphous and crystalline states of selected materials, different component geometries, and perfect amount and type of external stimulus, etc. GeTe is a promising candidate for RF switching applications because of its suitable and small resistivity in crystalline phase. Uniform heating-based phase transition is needed to attain RF switching which provides massive resistance change. Phase transition done by voltage pulses can provide the above advantage. When current pulse is provided for long time, induced heat goes above the crystallization temperature and recrystallization starts in GeTe films. On the other hand, amorphous phase returns in GeTe films when an increased current pulse is provided for shorter time [2, 3].

Even though terahertz (THz) technology can play vital role in several research fields, i.e. communication, biomedical, security, etc., it is often overlooked as there is little to no simple methods available for production, detection, and modulation of THz waves. Nevertheless, there are some technologies available which can be exploited to utilize THz waves. Among those, technologies like photonic crystals, quantum cascade lasers, and metamaterials are considered to be most promising. GeTe-based resonators and modulators can be applied in these applications [4, 5]. Besides switching and THz applications, GeTe is applicable in reconfigurable devices. These devices have the ability to be used for multiple applications which are very useful and cost-effective for circuit and other applications. For a basic RC circuit, by tuning the resistivity and permittivity of phase change materials, cutoff frequency can be changed because cut-off frequency is directly related to resistance and capacitance. Resistance can be changed by controlling the geometries and chemical structures of phase change materials-based resistors fabricated within the circuits. This change can alter the state of reconfigurable devices [6]. This chapter will discuss the applicability of GeTe in these three applications.

2. DC and RF switching applications

2.1 GeTe strips for direct current (DC) switching

Here, a new DC switch design is presented where no mechanical component is required to control the current flow in GeTe wire. Micromachined parallel wire designs were fabricated to apply external electric fields across the GeTe wire. First, oxide layer was created on Si substrate. GeTe wires/strips were patterned and deposited using RF sputtering method. Traditional lift-off method provided perfect GeTe wire shape. After that, gold contact pads (using Ti adhesion layer underneath) were patterned and deposited on top of GeTe wire using evaporation method. These GeTe wires were around 200 nm thick. Using the similar method, parallel gold wires/plates were patterned and deposited around the GeTe wire. These parallel plates were used to provide electric fields. Gap between the wires were 80 and 100 μm. **Figure 1** shows the design of GeTe test structure [1].

Resistivity and switching property of GeTe are highly dependent on the deposition condition. Upon application of an external stimuli, an amorphous insulating GeTe film transitions into conducting polycrystalline. As the GeTe strip is placed in between two parallel plates, electric field (E) across the device, that is generated by adding bias to the plates, can be found from the applied voltage (V) across the parallel wires and the distance (d) between them (shown in Eq. (1)). GeTe strip resistance can be determined according to the Ohm's law. To prove the phase transition, GeTe resistivity ($\tilde{n}$) is calculated using the measured strip resistance (R) and the length (l) and cross-sectional area (A) of GeTe strip (shown in Eq. (2)). Any drastic change of resistivity due to electric field implies a phase change in the GeTe wire [1]:

$$E = \frac{V}{d} \tag{1}$$

DOI: http://dx.doi.org/10.5772/intechopen.108461

$$R = \frac{\rho l}{A} \tag{2}$$

Electric field was created across the GeTe wires by applying voltage across the parallel plates and resistivity of GeTe wires changed. So, transition from conductor phase to insulator phase occurred. This applied electric field becomes higher and lower, respectively, when voltage is gradually increased and decreased. DC probes were connected with the gold contact pads (test pads) which gave the resistance values. Increase in resistivity was found about three to five orders of magnitude, depending on the different lengths and widths of GeTe wires. It was also found that GeTe wires, while in the volatile regime, regained their initial material phase when the external voltage was eliminated. **Figure 2** shows the resistivity variation according to different applied voltages and different sizes of GeTe wires. The transition happened when the applied voltage was in between 37 to 45 V [1].

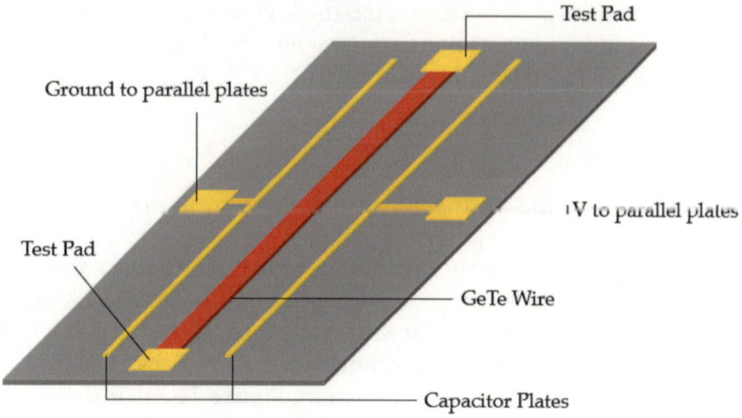

Figure 1.
Design of GeTe test structure for DC switching testing [1].

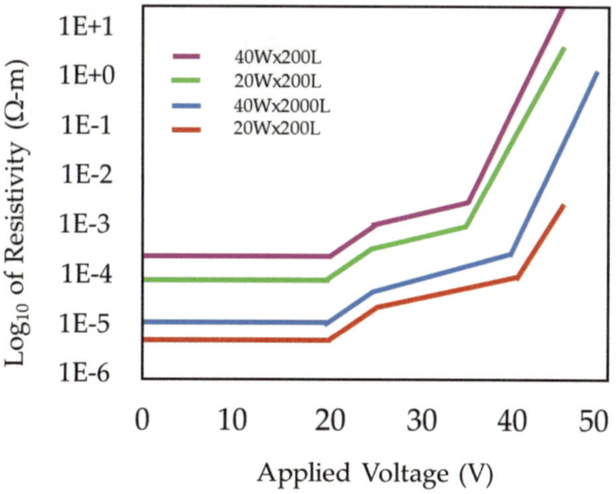

Figure 2.
Resistivity variation in GeTe wires because of external electric field [1].

2.2 GeTe-based coplanar waveguide (CPW) radio frequency (RF) switches

2.2.1 Fabrication process for GeTe RF switches

For this fabrication process, oxide and nitride layers were used as isolation layers. Some substrates had oxide layer, and some have nitride layers. Silicon substrate was thermally oxidized to make silicon dioxide (300 nm thickness) film. On the other hand, plasma-enhanced chemical vapor deposition (PECVD) was used to make silicon nitride (100 nm thickness) film. On top of these isolation layers, GeTe strips for signal line were deposited. GeTe layers were fabricated using RF sputtering method (parameters were 300 W RF power, 10 mTorr pressure, 20.1 sccm Ar gas flow for ignition, and 2 min time). The prepared GeTe films were in amorphous phase because of room temperature deposition. GeTe strips had length of 30 μm, width of 10 μm, and thickness of 100 nm. After that, bimetallic Ti/Au ground lines and bimetallic Ti/Au pads on both sides of GeTe strips were deposited using E-beam evaporation method and lift-off method. Ten-nanometer-thick Ti works as an adhesion layer for 100 nm gold layer, and it is essential to get better electrical contact. **Figure 3** shows the fabricated GeTe CPW RF switching devices [2].

2.2.2 Performance analysis of GeTe RF switches

Before doing RF switching testing, thermal characterization was done. Thermal characterizations were done using thermal chuck where devices were directly placed on the chuck. **Figure 4** shows the thermal characterization results, and it is found that all devices show resistance change of ~six orders of magnitude. **Figure 4a** and **b** shows nonvolatile phase transitions. These devices were fabricated using oxide and nitride isolation layer, and GeTe film resistance never came back to its original values. This is an indication of permanent crystallization. **Figure 4c** and **d** shows volatile phase transitions. These devices were directly fabricated on substrates (c-Si and Al_2O_3 substrates, respectively) without any isolation layer, and they came back to their original values. Also, **Figure 4c** does not show ideal I-V behavior because of higher conductivity of c-Si substrate [2].

This heat-induced phase transition can be explained by the phenomena of heat transfer via conduction through multiple layers of materials. For GeTe on

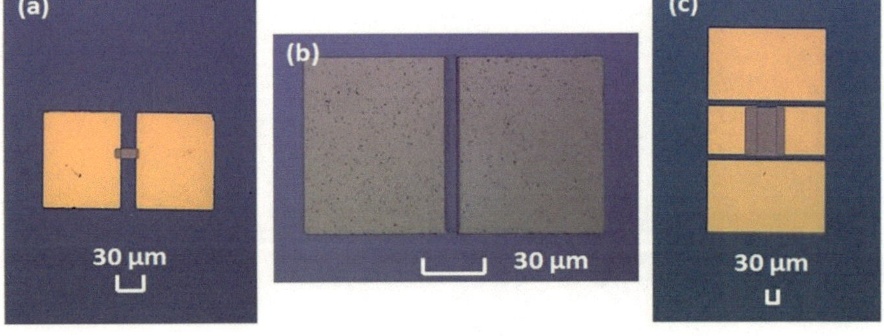

Figure 3.
(a) Microscopic image of a device with GeTe strip; (b) a smaller device on sapphire substrate for I-V measurement; and (c) a CPW RF switch having signal and ground lines [2].

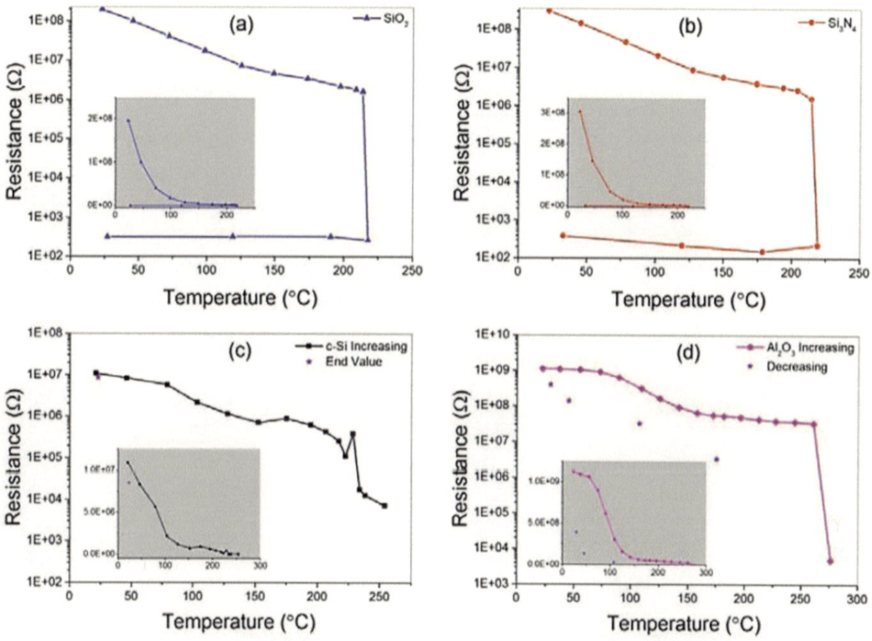

Figure 4.
Semi-logarithmic resistance versus temperature curves of GeTe: (a) on SiO_2/c-Si substrate; (b) on Si_3N_4/c-Si substrate; (c) on c-Si substrate; and (d) on Al_2O_3 substrate (LINEAR scale curves are in the insets) [2].

SiO_2/c-Si and GeTe on Si_3N_4/c-Si, there were two layers of materials in between GeTe layer and thermal chuck. For GeTe on c-Si and GeTe on Al_2O_3, there was only one layer of material in between GeTe layer and thermal chuck. Heat vertically moves upward direction from thermal chuck. Small crystallites are formed due to this upward flow from substrate to GeTe via dielectric layer, and these crystallites are responsible for the GeTe resistance change. Additionally, GeTe on Al_2O_3 shows volatile transition following a different path. This is because of the difference in thermal expansion coefficients between Al_2O_3 and GeTe (5.3 and 0.56, respectively), which may result breaking of the conductive crystallites throughout the cooling process [2].

To perform S-parameters analysis of GeTe-based RF switches, S_{11} and S_{21} were investigated in frequency range of 10 MHz-20 GHz. It was found that when frequency increased, S_{11} went down from −11 dB to −16 dB in crystalline phase. In crystalline phase, reflectivity increases because of incorporation of small crystallites due to frequency increase. In amorphous phase, full return loss was found as S_{11} remained constant according to frequency change. These are shown in **Figure 5**. S_{21} was found constant in crystalline phase, and the value was −3 dB. In amorphous phase, S_{21} had different values in different frequency range. In 10–100 MHz range, it was −55 dB and remained constant. But when frequency increases from 100 MHz to 20 GHz, S_{21} gradually increased, and the value went up to −28 dB. This might be because of frequency induced crystallization in amorphous phase. The signal line resistance was found as 75 Ω and 3.5 MΩ in ON and OFF states, respectively. For making high-performance RF switch, it was suggested that S_{21} losses can be reduced by reducing the width of GeTe section which can eventually reduce ON state resistance from 75 Ω to 50 Ω [2].

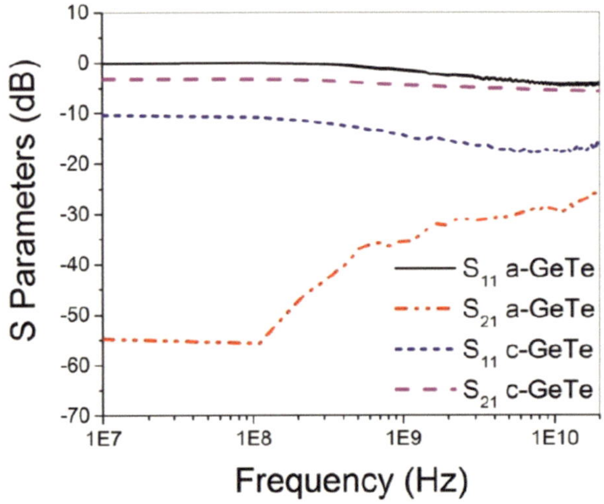

Figure 5.
Analysis of S parameters for GeTe-based RF switches. In this analysis, 31.6% and 10% of full signal strength are equivalent to −10 dB and −20 dB [2].

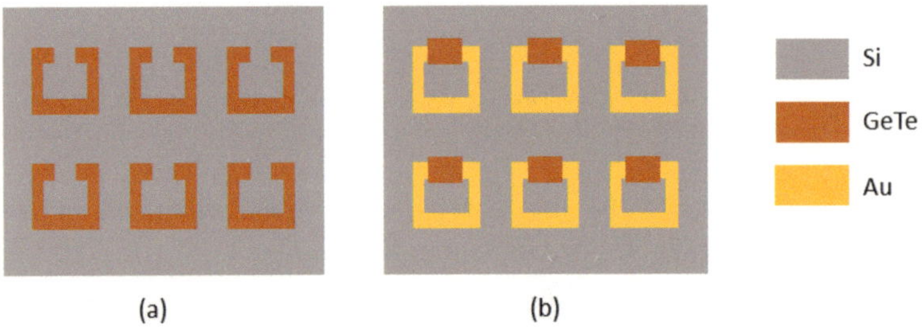

Figure 6.
GeTe gold split-ring resonators (SRRs): (a) GeTe SRRs; (b) GeTe-in-gap SRRs.

3. Applications in terahertz (THz) technologies

3.1 GeTe-based THz split ring resonators

This section will focus on metamaterials compatible with CMOS technology and MEMS fabrication processes. A terahertz (THz) split-ring resonator (SRR) made from germanium telluride (GeTe) has been developed. This metamaterial was chosen for creating tunable response. Approximately at 180–230°C, amorphous GeTe becomes crystalline. The precise phase changing temperature depends on GeTe film thickness, its stoichiometry, and the substrate material. The SRRs here have a 5 μm line width, 20 μm square side length, 3 μm gap width, and a 39 μm periodicity between elements. 300-nm-thick GeTe films were sputter deposited on these devices using 99.99% pure, 50/50 GeTe sputter target (**Figure 6a**). The sputter chamber was at 10 mT with 20.1 sccm Ar gas flow. 300 nm Au deposition was done for another batch of devices using electron beam evaporator system (**Figure 6b**). The deposited metals were patterned in bilayer lift-off method [4].

At 180–190°C, both types of devices showed abrupt change in transmission line and behaved as metal, indicating GeTe's crystalline phase. Overall, GeTe SRRs showed transition amplitude drop (**Figure 7**), whereas GeTe-in-gap SRRs showed altered transmission shape and that did not change even after cooling because of their nonvolatile nature (**Figure 8**). At temperatures above crystallization temperature, transmission kept falling. Even though it is known that GeTe gets less conductive at higher temperatures, it is observed that GeTe got more conductive along with increasing temperatures. This is because of the presence of thermally induced free carriers at the silicon substrate [4] (**Figure 9**).

Although it was expected to get sharp resonances in the THz transmission response of the GeTe SRRs beyond crystallization temperature, it was not observed. This absence of resonances can be explained by considering ohmic losses in the crystallized GeTe. Therefore, it was found that at THz speeds, c-GeTe's conductivity nature is not as ideal as gold. In case of in GeTe-in-gap SRRs, it was noticed significant change in transmission response after crystallization. Unlike the GeTe SRRs, impact of ohmic losses was not so daunting. This is for the fact that the devices had Au in mostly with a comparatively small portion of GeTe in the middle. Based on the outcome of GeTe-in-gap SRRs, it can be concluded that GeTe can be successfully utilized along with metals for improving tunability of such devices. GeTe-incorporated metamaterials can be applied in modulation by adding electrical GeTe switching circuits or by optimizing the GeTe film parameters for optical switching [4].

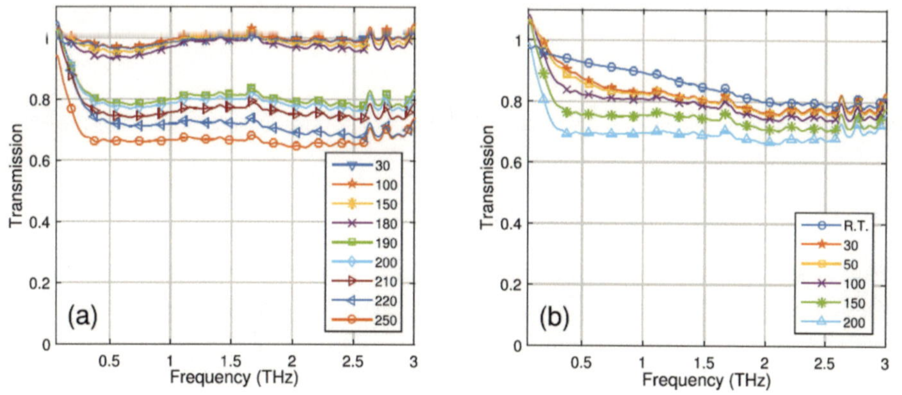

Figure 7.
GeTe SRRs transmission responses for (a) increased and (b) decreased temperature [4].

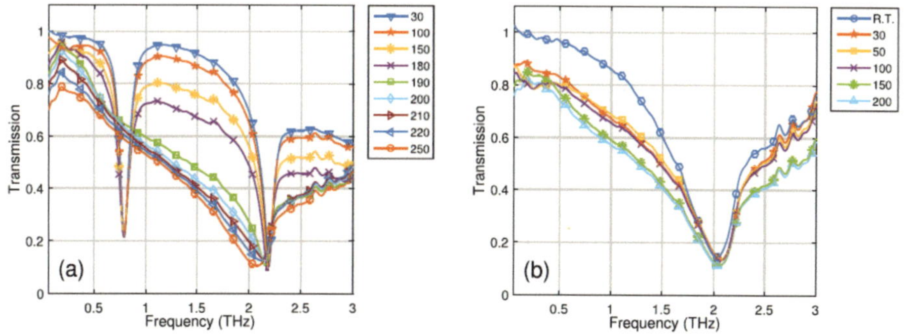

Figure 8.
GeTe-in-gap SRRs transmission responses for (a) increased and (b) decreased temperature [4].

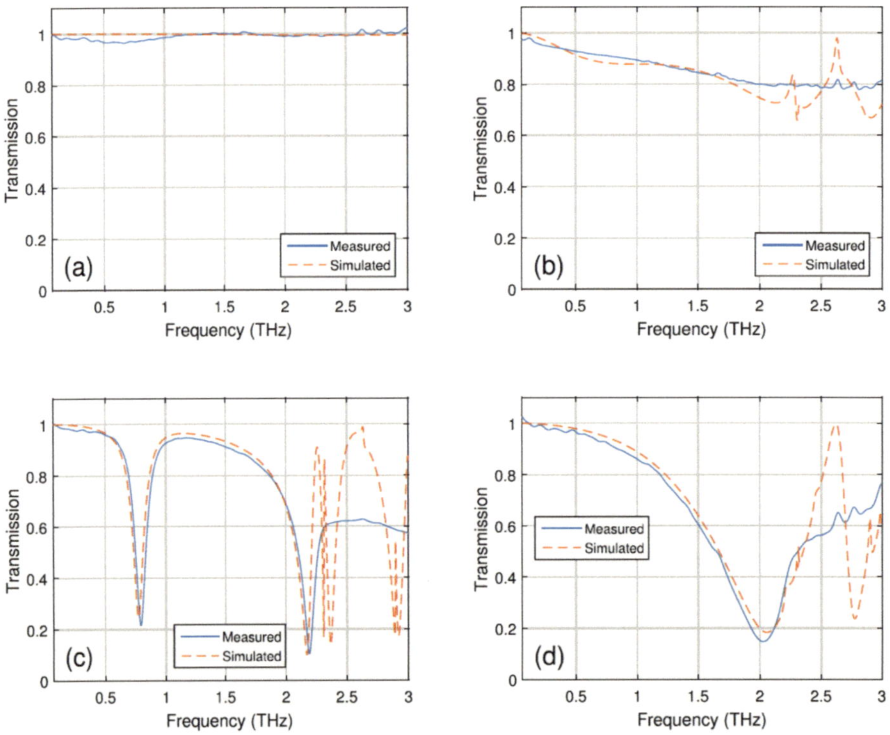

Figure 9.
Comparison of simulated and measured data: (a) a-GeTe SRRs; (b) c-GeTe SRRs; (c) a-GeTe-in-gap SRR, and (d) c-GeTe-in-gap SRR [4].

3.2 Improved THz modulations using GeTe

Here, a study has been done on exploiting a single GeTe layer as a thin film THz modulator. A 100-nm-thick GeTe layer was deposited in same manner as it has been discussed in the previous subsection. Apart from that, a sapphire wafer was used as the control sample for the reference measurements. The reason behind choosing sapphire is because its rhombohedral elementary cell is comparable to c-GeTe. Following **Figure 10** shows time-dependent transmitted THz signals. It is evident from the figure that signal attenuation is highest around crystallization temperature. Here, rising of signal peak attenuation and time delay between signal generation and detection have been seen. These changes are directly correlated with increased absorbance and refractive index, respectively [5].

Figure 11a shows that the absorbance increases as the temperature is increased. Highest absorbance was observed at the transition/crystallization temperature. Also, high absorbance region starts at 100 °C from 70 to 100 cm^{-1}. This region corresponds to incident radiation above 2 THz. Below transition temperature film stays in polycrystalline phase and acts as insulator. Region for below transition temperature phonon vibrations do not last long as the films are thin with pinholes and other film anomalies compared to substrate beneath. As a result, the absorbance was low. **Figure 11b** shows that without sapphire substrate's influence, for the GeTe only samples, around 88.5% to 91.5% modulation depths from crystalline GeTe were found. However, modulation of 99% at approximately 77 cm^{-1} (2.3 THz) is observed when c-GeTe/sapphire heated to 250°C. For our case, GeTe only samples

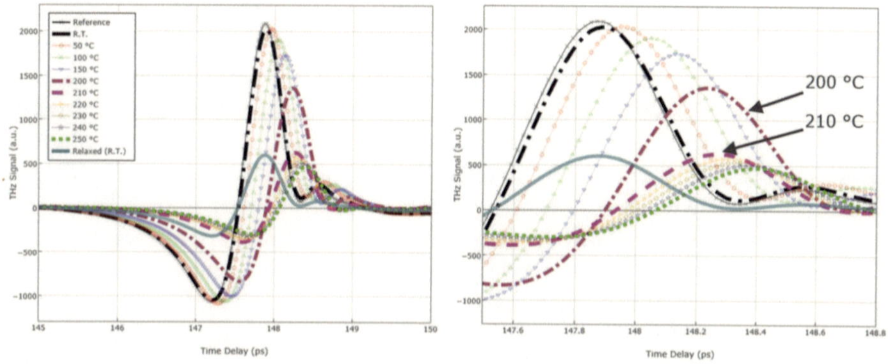

(a) (b)

Figure 10.
THz-TDS response of GeTe film from room temperature to 250 °C: (a) complete response of positive and negative time domain peaks; (b) close up view of positive time domain peak [5].

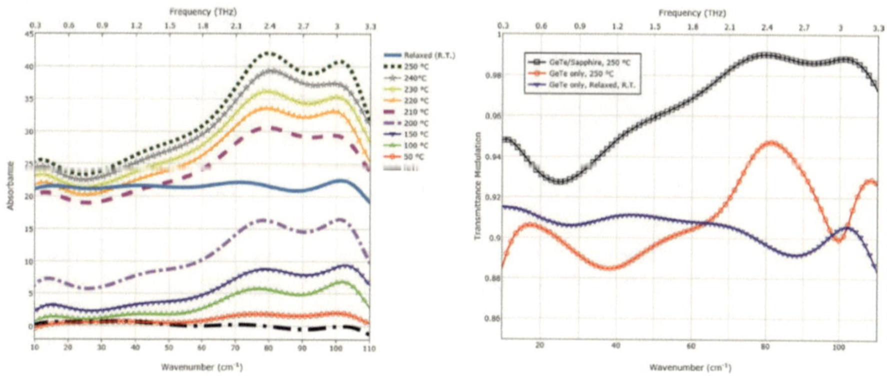

(a) (b)

Figure 11.
(a) GeTe film absorbance spectra with reference to sapphire substrate at room temperature to 250°C and again cooled to room temperature, across the full spectrum 10–110 cm^{-1}; (b) THz transmittance modulation of GeTe/sapphire and GeTe at 250°C and GeTe at room temperature [5].

are important for THz applications as their modulation does not fall below 90% up to 77 cm^{-1} (2.3 THz) [5].

4. Reconfigurable circuit applications

4.1 Fabrication of horizontal and vertical GeTe resistors

Two different geometries were chosen here. One is vertical GeTe resistor and other is horizontal GeTe resistor. In both horizontal and vertical resistors, there are bottom and top metal layers, GeTe layer and an isolation layer. To fabricate the horizontal resistor (**Figure 12**), firstly, a bottom metal layer (100 nm Au on top of 10 nm Ti) was patterned and deposited on silicon wafer using photolithography, e-beam evaporation, and traditional lift-off methods. Heat dissipation was happened via this bottom thermal conduction layer which is very important to maintain uniform crystallization.

Next, an isolation layer consisting of 100 nm silicon nitride (Si_3N_4) layer was prepared using plasma-enhanced chemical vapor deposition (PECVD) method. This layer created an isolation in between bottom metal layer and top contact pads. On top of the isolation layer, 100–200 nm thin GeTe line resistor (shape of narrow bridge) was deposited using high vacuum RF sputtering method in room temperature. Then, top metal contact pads (250 nm Au on top of 10 nm Ti) were patterned and deposited on both sides of GeTe resistor. In these horizontal resistors, inter-electrode distances were ranged from 5 μm to 10 μm. In these horizontal resistors, resistor length was represented by separation between two metal pads and resistor area represented by the product of GeTe thickness with distance between metal pads [6].

In the vertical resistor (**Figure 13**), Ti/Au bottom contact pads were patterned and deposited on Si wafer. This process was similar to the process of bottom metal thermal conduction layer of horizontal resistor, but this bottom contact pad was the bottom electrode. 100 nm Si_3N_4 isolation layer was prepared on top of it. Next, hole was patterned and created in the Si_3N_4 isolation layer using photolithography and reactive ion etching (RIE) methods. Via this hole, GeTe with desired thickness was deposited using RF sputtering method, and this GeTe film connected the top and bottom electrodes. Then, Ti/Au top metal contact pads were patterned and deposited on top of GeTe which fully covered the GeTe film. In these vertical resistors, resistor length was represented by GeTe thickness and resistor area represented by the size of via hole [6].

4.2 Thermal and electrical characterization of GeTe resistors

Impact of direct heating (Joule) and indirect heating (thermal) and the relation between geometrical parameters and threshold voltage have been revealed. First, amorphous phase resistance was investigated for both horizontal and vertical GeTe resistors according to their thickness (**Figure 14a**). It was found that the resistivity of vertical resistors was hugely dependent on layer thickness of GeTe. 100 nm and 400 nm thick vertical resistors gave resistance values of $1 \times 10^4\ \Omega$ and $8 \times 10^4\ \Omega$, respectively. Again, horizontal resistors showed greater resistance values than vertical resistors having same thickness. It was because of their larger inter-electrode distance

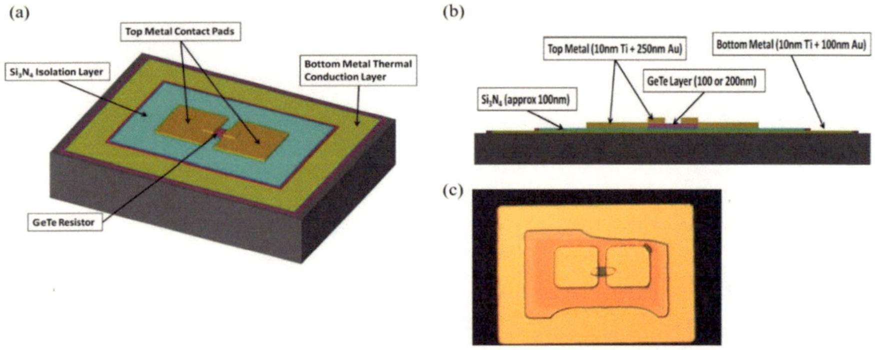

Figure 12.
(a) Top view; (b) cross-sectional view of GeTe horizontal resistor; (c) a GeTe horizontal resistor with 5 μm inter-electrode distance [6].

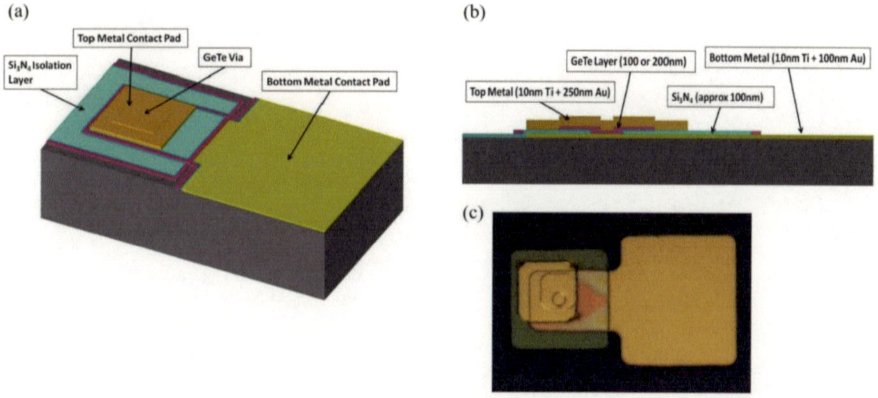

Figure 13.
(a) Top view; (b) cross-sectional view of GeTe vertical resistor; (c) a GeTe vertical resistor with 3 μm hole diameter [6].

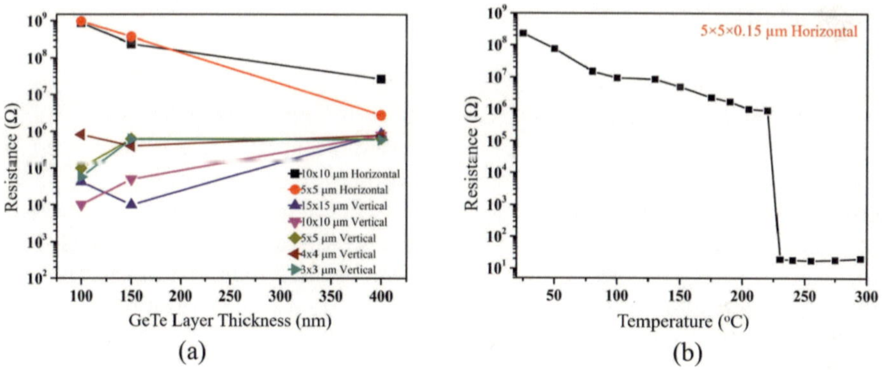

Figure 14.
(a) Resistance of different GeTe resistors in amorphous phase, and (b) phase transition of GeTe horizontal resistor using thermal conduction method [6].

and smaller cross-sectional area. It was also found that resistors having same device volume and different device geometry showed different resistance values [6].

To understand the amorphous to crystalline transitional behavior of GeTe resistors via thermal conduction method, temperature of thermal chuck was increased up to 300°C (**Figure 14b**). Transition in horizontal resistors was happened at 220–230°C where resistance change was found at MΩ range (a sudden drop from MΩ to Ω). On the other hand, in Joule heating-based phase transition, vertical resistors with different thickness and same hole area were used. As thickness increased from 90 nm to 400 nm, voltage for transition increased from 0.5 V to 3.25 V (**Figure 15a**) and threshold field also increased from 3.3 V/μm to 8.13 V/μm. For horizontal resistors, threshold voltage and threshold field are much higher because of larger inter-electrode distance (with same cross-sectional area). It was also established that the effect of cross-sectional area on threshold voltage is very negligible (**Figure 15b**). Finally, it can be concluded that despite having relatively complex fabrication process than horizontal resistors, vertical resistors can be fit for low-power reconfigurable circuits applications because of their smaller inter-electrode distance [6].

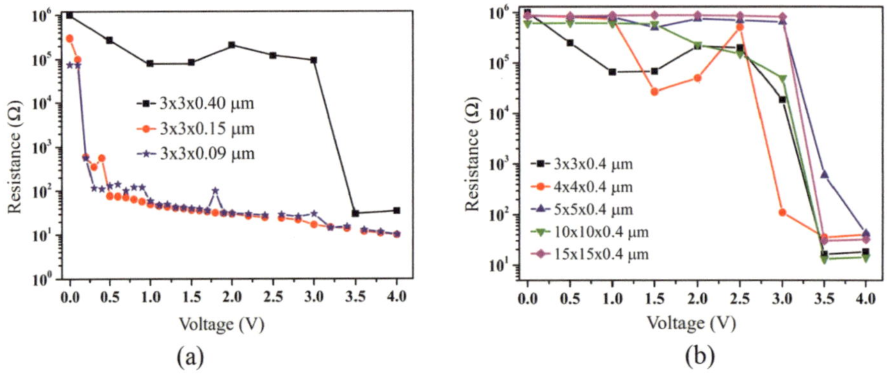

Figure 15.
Joule heating-based transition of GeTe vertical resistors: (a) effect of inter-electrode distance with same cross-sectional area and (b) effect of cross-sectional area with same inter-electrode distance [6].

5. Conclusion

This book chapter discusses about how it is possible to exploit a chalcogenide phase change material like GeTe for DC and RF switching, THz technology, and reconfigurable circuits applications. A GeTe-based test structure was fabricated for DC switching testing. It was found that phase transition occurs when at 37 to 45 V applied voltages. More importantly, GeTe wire was able to go back to its original phase when the voltage was removed. A PCM device was fabricated for analyzing GeTe's performance for RF switching applications. It was evident from the results that a thinner GeTe section in the device can reduce ON state resistance and hence improve RF switching performance. Following that, different types of GeTe-based SRRs were fabricated for applications related to terahertz (THz) technologies. It is found that, for certain device designs and GeTe placements, GeTe can provide significant tunability of metamaterial devices. For a GeTe/sapphire system, 92%–99% modulation depths were achieved. By observing the modulation depth results, it can be assumed that incorporating GeTe into modulators, filters, and other THz technology devices will be beneficial for current THz technology. Moreover, from the experimental results found from GeTe resistors, it can be concluded that even though vertical resistors are harder to fabricate, they are suitable for low-power reconfigurable circuits applications as they have smaller inter-electrode distance. Based on all the experiments discussed in the book chapter, it is justified to say that GeTe is a perfect candidate for representing the positive impact of chalcogenide phase change material on current THz technology field.

Acknowledgements

The authors want to thank all the authors and co-authors whose works have been discussed in this book chapter.

Author details

Turja Nandy, Farhana Anwar and Ronald A. Coutu Jr.*
Marquette University, Milwaukee, United States

*Address all correspondence to: ronald.coutu@marquette.edu

References

[1] Danner BL, Coutu RA Jr. Characterizing metal-insulator-transition (MIT) phase change materials (PCM) for RF and DC micro-switching elements. Procedia Engineering. 2012;**47**:80-83

[2] Gwin AH, Coutu RA. Electronic control of germanium telluride (GeTe) phase transition for electronic memory applications [Internet]. In: Teherani FH, Look DC, Rogers DJ, editors. Oxide-based Materials and Devices VI. SPIE; 2015. Available from: http://dx.doi.org/10.1117/12.2079359

[3] Mahanta P, Munna M, Coutu RA. Performance comparison of phase change materials and metal-insulator transition materials for direct current and radio frequency switching applications. Technologies. 2018;**6**(2):48

[4] Kodama CH, Coutu RA Jr. Tunable split-ring resonators using germanium telluride. Applied Physics Letters. 2016;**108**(23):231901

[5] Gwin AH, Kodama CH, Laurvick TV, Coutu RA Jr, Taday PF. Improved terahertz modulation using germanium telluride (GeTe) chalcogenide thin films. Applied Physics Letters. 2015;**107**(3):031904

[6] Sattler J, Coutu RA. Germanium Telluride (GeTe) phase change resistors for reconfigurable circuit applications. In: Proceedings of the 2015 National Aerospace and Electronics Conference (NAECON); 15-19 June 2015. Dayton, Ohio: IEEE; 2016. pp. 224-227

Chapter 9

Photo-Induced Displacive Phase Transition in Two-dimensional $MoTe_2$ from First-Principle Calculations

Yiming Zhang, Yuanfeng Xu, Yujie Xia, Juan Zhang, Hao Zhang and Desheng Fu

Abstract

The discovery and control of new phases of matter are a central endeavor in materials research. Phase transition in two-dimensional (2D) materials has been achieved through laser irradiation, strain engineering, electrostatic doping, and controlled chemical vapor deposition growth, and laser irradiation is considered as a fast and clean technique for triggering phase transition. By using first-principles calculations, we predict that the monolayer $MoTe_2$ exhibits a photo-induced phase transition (PIPT) from the semiconducting 2H phase to the topological 1T′ phase. The purely electronic excitations by photon soften multiple lattice vibrational modes and lead to structural symmetry breaking within sub-picosecond timescales, which is shorter than the timescale of a thermally driven phase transition, enabling a controllable phase transition by means of photons. This finding provides deep insight into the underlying physics of the phase transition in 2D transition-metal ditellurides and show an ultrafast phase-transition mechanism for manipulation of the topological properties of 2D systems. More importantly, our finding opens a new avenue to discover the new families of PIPT materials that are very limited at present but are essential to design the next generation of devices operated at ultrafast speed.

Keywords: photo induced phase transition, two-dimensional material, ab-initio calculations, electronic excitation, landau theory

1. Introduction

1.1 Current status of photo-induced phase transition (PIPT)

Phase transition is a process by which matter changes from one phase to another. The discovery, characterization, and control of materials at different phase are a central endeavor in condensed matter physics and materials science [1]. The traditional methods using theoretical and experimental methods including thermal annealing, strain engineering, charge doping, adsorption, alloying and electron injection have been adopted to understand the transformation of phase transition [2]. However, the abovementioned phase transformation states are all stabilized in the so-called equilibrium phases, and some require a complex environment to generate the phase transition. Meanwhile, with the increasing importance of optical modulation in the aspect of modulating materials structures and properties in recent years, PIPT has become an important tool in adjusting materials in the aspect that the abovementioned traditional methods are ineffective [2].

PIPT is a phase transition directly by light that can be precisely controlled toward target structures with desired physical properties at transient speeds, which is much more convenient than using heat or electric field to control the phase transition. The process of phase transition is basically a non-equilibrium process, and it is especially important to understand the evolution process of this non-equilibrium phase transition. Therefore, it is necessary and extremely important to obtain basic information such as timescale and driving force of PIPT. Using light to control ultrafine transformation between different forms of matter to realize ultrafast functional devices is the goal that scientific researchers have been pursuing since the discovery of PIPT in the 1990s, which takes place in one-dimensional (1D) corrected organic charge-transfer molecular system tetrathiafulvalene-p chloranil (TTF-CA) [3]. Actually, because light-induced charge transfer excitation and/or carrier generation can excite the instability of 1D properties inherent in electronic states through strong electron (spin)-electron (spin) interactions, researchers treat the PIPT of 1D correlated electron systems as their targets.

Although PIPT has the advantages of ultrafast dynamics and simple operation, the areas where PIPT phase transition can be successfully implemented are limited, such as: (1) time-resolved spectroscopy of the dynamics of photo-induced ionic-to-neutral or semiconductor-to-metal phase transition [4]: mixed-stack organic charge transfer crystal, color-phase transitions in polydiacetylene and polythiophene, for example. In such PIPT system, it has strong electron-lattice coupling during the cooperative relaxation interactions; (2) low-spin to high-spin state conversion related magnetic materials or light-induced full topological physics in magnetic semiconductors: $[Fe(2pic)_3]Cl_3 \cdot EtOH$ and RbMn $[Fe(CN)_6]$, for example [5, 6]. Phase transition in magnetic materials under the action of light is caused by the coupling between volume striction effect and acoustic modes, and the abrupt step transition observed in the experiment is attributed to the depinning effect of volume shrinkage; (3) femtosecond/picosecond dynamics of the photo-induced lattice rearrangements: $[Pt(en)_2]$ $[Pt(en)_2X_2](ClO_4)_4$ with X = Cl, Br and I, for example [4]. The formation and decay dynamics of self trapped excitons are the starting point of various metastable formation processes accompanied by lattice rearrangement as the excited states of electrons induced by photo-excitation change chemical bonds or generate metastable states; (4) PIPT also presented in ferroelectric materials: barium titanate ($BaTiO_3$) and lead titanate ($PbTiO_3$) in perovskite oxides system, for example [7]. With the typical property of polar distortions in ABO_3 perovskite oxides, the polarization within it nearly disappears under the stimulation of light, nonpolar phases such as antiferroelectric come into being with the tilt of the oxygen octahedra in ABO_3 perovskite oxides.

The emergence of 2D and the layered van der waals materials gives the research a new platform of the diffusive, displacive, and quantum phase transitions with distinct physical properties arised from the dimensionality confinement, chemistry, electrostatics, elasticity, and defects [8]. Recently, experimental evidence has shown that the laser irradiation induced phase transition observed in $MoTe_2$ and WTe_2 often along takes place with the abrupt changes of electronic structure, especially with the spring up of some new topological phases [9–11]. Taking few-layer $MoTe_2$ under laser irradiation, for example, it experiences an irreversible transition from the semiconductor hexagonal 2H phase to the topologically distorted octahedral 1T′ phase. And it can be used in the desired area for the fabrication of an ohmic heterophase homojunction with the accurate command of micrometer patterning. Despite the intensive investigations, the microscopic mechanism of the laser irradiation induced phase transition in these 2D materials is still not very clear. These experimental findings also suggest that 2D materials may provide a new platform to discover the PIPT materials that is the focus of this theoretical study.

1.2 Scientific importance of PIPT

Recently, experimental observation of ultrafast disordering of atomic motions in PIPTs has challenged the convertional knowledge of phase transition that the atomic motion in PIPTs is always treated as a coherent process [12]. Actually, ultrashort laser pulses manipulate the structure and function of materials, transforming the electronic or magnetic properties with a timescale in the limit of atomic motion and providing a powerful approach under photo-excitation. Active optical control or interplay between optical and other external influences impact matter or materials are desirable in the field of scientific disciplines, which may display fresh fundamental theoretical and device concepts by optically induced electron-electron correlations, electron-lattice coupling, altering band structures and topology, and so on [4]. Photoexcitation is not only widely used in one-dimensional material systems over the past three decades [13], but also emerging in other fields in recent years, such as memory devices, twistronics field, in-sensor computing field, ultrafast dynamics, and ultra-high frequency acoustic devices and so on. It's all due to the rich physics effects of PIPT.

Light can introduce the charge density wave as the phase competition in far-from-equilibrium system of $LaTe_3$ [14], with the presentation of topological defects under the influence of slower re-establishment of phase coherence [15]. This result paves the way for understanding the mechanical nature of topological defects with non-equilibrium defect-mediated transitions and providing a framework for the production, control, and manipulation of optically introduced other ordered phases. PIPT can also be used in the emerging in-sensor computing field [16, 17]. Under oxygen stoichiometry engineering, VO_2 films can show non-volatile multi-level control under ultraviolet irradiation, with integrated processing, sensing, and memory functions at 300 K. It can significantly improve the recognition accuracy from 24 to 93% because this artificial neural network can exact UV information from ambient environment consisting of the abovementioned neuromorphic sensor. This PIPT observed in the application of neuromorphic ultraviolet sensors not only provides an avenue of the design of neuromorphic sensors but also facilitates its potential applications in artificial vision systems.

Furthermore, in various field such as complex liquids, biological and sensing nanostructures systems, ultrafast light-generated giga–terahertz (GHz–THz) frequencies acoustic phonons have attracted much attention [18–20]. For example, ferroelectric $BiFeO_3$ has photogeneration/photodetection of coherent phonons [21], with the generation of strain under giant ultrafast light because of the generation of stress by inverse piezoelectric effect under the screening of the internal electric fields by light-induced charges. This giant opto-acoustic response opens up a new prospect for the application of ferroelectric oxides in ultra-high-frequency acoustic devices and promoting the development of some new GHz-THz sound sources. Meanwhile, the corresponding ultrafast dynamics of materials at far from equilibrium states under laser irradiation are one of the ultimate problems in modern science and technology [22, 23]. Actually, PIPT is always treated as a coherent process, but ultrafast disordering of atomic motions in PIPT is observed in recent experiments [24]. It is very important to understand the evolution process of the non-equilibrium phase transition and obtain the basic information such as timescale and driving force of the structural phase transition.

In fact, there are various opinions on the basic information of timescale and driving force in PIPT. The competitive relationship of all driving forces such as photo-excitation, thermal phonon vibration, strain, and so on, has been the focus of scientific attention. With low laser fluence, the timescale of phase transition is always long under the combination of the atomic driving forces caused by thermal phonon vibration and photo-excitation, due to influence of photo-excited hot carriers cooling. In VO_2, at high laser fluence, the smallest timescale of phase transition is indeed saturated to a minimum value of about 55 fs, which is due to the saturation of the population of the V–V bonding states by the photo-excited holes [24]. In this work, the structural transition takes place within 1 picosecond, shorter than the timescale for the photo-excited electrons to transfer their energy to the lattice. We also rule out a thermally driven or a strain-induced phase transition under laser irradiation by comparing the thermodynamic stability.

1.3 Novel two-dimensional PIPT material family predicted from ab-initio calculations

Although PIPT has some applications, its physical mechanism has not been clearly defined, and there is still no unified method to explain this physical phenomenon. The microscopic nature of optically driven phase transition is still unclear. With the changes of strain, lattice vibrational modes, chemical state, electronic excitation, and temperature in the studied system excited by light, the corresponding phase transition may be triggered by one of these factors or on a combination of them. It is of great physical importance to understand the dynamic process of PIPT-induced evolution. However, there are still many technical difficulties to understand the evolution of the ultrafast phase transition both experimentally and theoretically. In this work, we adopt a mature computational method: first-principles calculation, combining ultrafast dynamics and Landau theory to predict the emergence of PIPT in 2D materials for the first time, which opens a new direction for the future search of PIPT materials.

Several models have been proposed to explain this phase transition mechanism: (1) local phase transition induced by Te vacancies created by irradiation [25]: in the presence of nearby vacancies, the contraction of the migration energy makes the higher rate of larger defects accumulation than that expected from the isolated migration. Therefore, it acts as a seed for gathering more vacancies around the photo-induced Te vacancy region, producing nucleus arrangement, increasing the growth rate of 1T′ phase; (2) accumulated heat is a main driving force for the phase transition [26]: the local instantaneous heating from a laser leads to a certain number of Te vacancies, which significantly reduces the potential energy barrier between 1T′- and 2H-$MoTe_2$; (3) laser-induced thermal strain contributes to phase transition [27]: strain effect originating from thermal expansion can tune the electronic properties and induce a phase transition between 1T′- and 2H-$MoTe_2$; (4) the electronic excitation plays a critical role in the phase transition [2, 28, 29]: with the increase of excited charge carriers density, the energy barrier between 1T′ crystal and electronically excited 2H phase decreases monotonically and decreases to the thermal energy fluctuation range under the carrier concentration.

Due to electronically excited population inversion, a new distorted trigonal prismatic phase of 1T′ $MoTe_2$ was generated in previous work [29], which will significantly broaden the application spectrum of these materials. The discovery and control of this phase transition are expressively important to electron regulation science of 2D materials, and the characterization of PIPT is quite less explored with its uncertainty. It is necessary to correctly explain the phase transformation mechanism or the relationship between the magnitude of activation barriers and the potential energy surface in the excited state. Here, based on the first-principle calculations, the microscopic nature of PIPT was studied by a pure photo-induced excitation processes, not by thermal accumulation, defects, or thermal strain as previously assumed. In our calculations, available electrons are stimulated by photo-excitation energies of 1.58 eV (785 nm), 1.96 eV (633 m), 2.34 eV (532 nm), and 2.63 eV (473 nm) respectively. On the premise of unchanging the chemical composition of 2D materials, electrons are excited by light, and the electronic distribution state and lattice vibration mode in the crystal are fundamentally changed in the process of excitonic excitation, which leads to the significant change of chemical bond. Meanwhile, the appearance of ultrafast phase transition to 1T′ $MoTe_2$ driven by lattice vibration mode softening when the photon energy is larger than 1.96 eV for 2H $MoTe_2$, and this phase transition by electron excited state is very similar to Peierls phase transition.

In this chapter, we not only demonstrate that the phase transition of monolayer $MoTe_2$ can be triggered by photo-excitation of carriers alone, but also successfully using Landau theory for the first time to reveal the difference between the physical mechanism of PIPT and the traditional equilibrium thermodynamic phase transition caused by temperature or pressure. The starting point of Landau's theory is that the symmetry of thermodynamic functions should be the same as the symmetry of the crystal structure of the system. The key point of Landau theory is that symmetry is broken by mean-field approximation or saddle-point approximation and the phase transition with non-zero order sign is obtained. The theory of Landau phase transition is based on the construction of Landau Free Energy after selecting the state variables corresponding to the order parameters of phase transition [30]. For a soft-mode displacive phase transition, the order parameter can be chosen as the amplitude of the distortion of the soft-mode eigenvector u [31]. In this work, we applied the Landau phase transition theory to the explanation of PIPT, the lattice vibration barrier of the E "and A2" uses Landau expansion, which gives the relationship between vibration mode frequency and excitation photon energy. Finally, the vibration mode frequency will drop to zero with the lattice vibration mode completely softened when the excitation energy reached 1.96 eV, which leads to the emergence of the structural phase transition. Our findings not only reveal the microscopic origin of PIPT 2D transition metal tellurides but also provide hope to motivate both fundamental and applied studies of ultrafast phase transitions in these new class of materials for topological switching and neuromorphic computing.

2. Numerical method: First-principle calculations

We use Vienna ab-initio simulation package (VASP) to calculate the ground electron states and phonon properties and stimulate the photo-excitation processes [32].

For the pseudopotentials, we use the PBE potentials of $Mo(4s^2 4p^6 5s^1 4d^6)$ and Te $(5s^2 5p^4)$ atoms [33]. The structural optimization, self-consistent field calculation are conducted under energy cutoff of 500 eV, kmesh of $19 \times 19 \times 1$, energy threshold of $10^{-6} eV$, Hellman-Feynman force threshold of $10^{-4} eV/\text{Å}$.

In the stimulation of optical excited electrons, we neglect the electron-phonon scattering process and all the thermal effects. We approximately assume that the distribution function of electron states is by quasi-Fermi-Dirac function with infinite smearing, thus the distribution becomes stepwise, as follows:

$$f(\varepsilon) = \begin{cases} 1, & \varepsilon < \varepsilon_{FV} \\ 0, & \varepsilon > \varepsilon_{FC} \end{cases} \tag{1}$$

During the excited state simulation, we manually lift and lower the quasi-Fermi levels by,

$$f(\varepsilon) = \begin{cases} 1, & \varepsilon < \varepsilon_{FV} \\ 0, & \varepsilon_{FV} < \varepsilon < \varepsilon_F \\ 1, & \varepsilon_F < \varepsilon < \varepsilon_{FC} \\ 0, & \varepsilon > \varepsilon_{FC} \end{cases} \tag{2}$$

The energy difference between the quasi-Fermi level ε_{FC} for electrons and the quasi-Fermi level ε_{FV} for holes is used as a merit of photo-excitation, which takes five discrete values, 1.58, 1.96, 2.34, and 2.63 eV respectively.

3. Results and discussions: PIPT in 2D $MoTe_2$

Firstly, we calculate the phonon spectra of high-symmetry hexagonal 2H phase and low-symmetry monoclinic 1 T′ phase. The results are depicted in **Figure 1** and fit well with previous reports [34]. The color from light yellow to dark red indicates the Bose-Einstein distribution of phonon states calculated at 300 K. Since the 2H phase has D_{3h} point group at Γ point, the vibration modes can be decomposed by irreducible representations as [35]

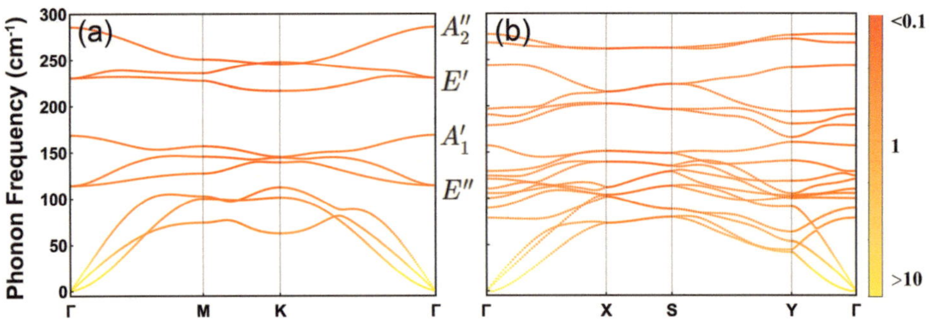

Figure 1.
Phonon dispersion of (a) high-symmetry 2H and (b) low-symmetry 1 T′ $MoTe_2$ monolayer and irreducible representation for optical phonons at the Γ point. The occupation number (calculated by Bose-Einstein distribution at 300 K) is marked by the colormap.

$$\Gamma_{2H} = E'' + A_1' + E' + A_2'' \tag{3}$$

We find that the E', E'', A_2'' phonons display well correspondance to the character of 2H to 1 T′ phase transition. The microscopic atomic displacements are shown in **Figure 2**(**d-f**) by arrows. Under photo-excitation by photons with energies of 1.58, 1.96, 2.34, and 2.63 eV, we calculate the one dimensional potential energy surface (PES) along these vibrations to figure out the responsible modes of phase transition. The results are summarized in **Figure 2**(**a-c**). The 0 eV photo-excitation represents the displacement in dark environment for reference. Without photo-excitation, there is an energy barrier of 80 meV between 2H phase and 1T' phase. As the optical excitation increases, the origin energy minimum of 2H phase bifurcates and the newly emergent energy minima correspond to 1T′ phase [2, 29].

The upper and lower layers of Te atoms move in the monolayer plane along opposite directions in E" mode (**Figure 2**(**d**)). In the dark environment, the Te atoms do not experience any damping force that drives into another phase. With increasing portion of excited electron states, the potential energy surface becomes flattened and finally forms a lineshape of Mexican hat at critical photon energy of 1.96 eV. This indicates a frozen soft mode that leads to a displacive phase transition. For the A_2" phonon, both Mo atoms layer in the middle and Te atom layer at the side move vertically in different directions (**Figure 2**(**f**)). Similar with the foregoing circumstance of E" phonon, a Mexican hat lineshape also forms if the photon energy is larger than the critical one. But the E' mode shows totally different behaviors under photo-excitation (**Figure 2**(**e**)). For E' mode, both Mo atom layer and Te atom layers move in the monolayer plane, but along opposite directions. The one-dimensional potential surface of E' phonon shows asymmetric feature where the right-hand-side energy maximum gradually lowers down and finally becomes an valley at 2.63 eV of photon-excitation, which indicates that a new metastable phase emerges. We notice that for the photons with energy between 1.96 and 2.63 eV, the energy maximum does not become an energy minimum, thus the E' does not anticipate the phase transition process and only E" and A_2" phonons make a contribution. After 2.63 eV excitation where the energy minimum along E' mode emerges, the E' mode starts to make a contribution simultaneously.

Besides the total energy surface, we can justify the stability of a phase by inspecting the phonon spectra. Since new energy minima emerge from a parabolic

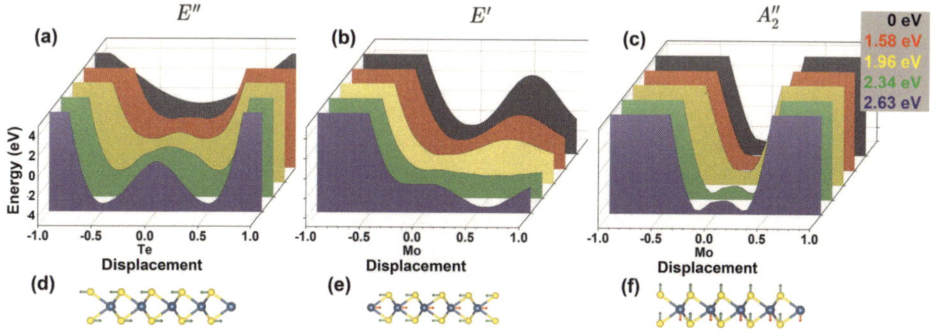

Figure 2.
One-dimensional potential energy surface at different excitation energies in the eigenspaces of (a) E", (b) E', and (c) A_2" phonons, and their corresponding structures (d–f) at the minima of the potential energy surface at 2.63 eV.

potential, which reveal new metaphases, the phonon spectra are also expected to collap and become softened. The frozen phonon mode can lead to structural phase transition, which breaks the original high symmetry [31, 36]. The E" and A_2" phonon frequencies are investigated using density functional perturbation theory under different photon excitations. The results are shown in **Figure 3**. The phonon frequencies of E" and A_2" phonon become imaginary as expected, which indicates a finite damping force and energy decreasing along the phonon mode. The critical photon energy fits well with the one of previous potential energy surfaces.

In the general Landau framework, we can choose an order parameter, which becomes finite at the vicinity of phase transition region, and the Landau free energy is expanded as a fourth-order polynomial of it, as follows:

$$F(u) = F_0 + au^2 + bu^4 \tag{4}$$

The coefficients a,b are related with the external condition. The change of sign of a can lead to an emergence of new energy minimum at $u\propto\sqrt{-\frac{a}{2b}}$. By choosing here the photon energy as an external condition indication and the phonon frequencies as the corresponding order parameter. We fit the frequency of E" and A_2" modes against the photo-excitation energy and obtain good agreement given by $\omega\propto(E_C - E)^{\frac{\gamma}{2}}$, where E_C = 1.96 eV and $\gamma = 0.24$, as shown in **Figure 3**. The result is similar as the well-known Curie-Weiss law, except that the critical exponent is 0.12. We can call it as the modified Cuire-Weiss law for PIPT. The sudden softening of E" and A_2" phonon modes further proves the PIPT.

The high-dimensional potential energy surface (PES) in the configuration space as a direct product of E" and A_2" modes under different photo-excitation is depicted in **Figure 4(a–d)**. The photo-excitation is chosen as 1.0, 2.0, 3.0, 4.0% per unit cell. These results suggest that at ~ 3.0% excitation, the energy minimum bifurcates and two new asymmetric PES minima emerge. Besides, in comparison to the E" mode, the A_2" mode apparently contributes more, which indicates that the phase transition is mostly triggered by A_2" mode.

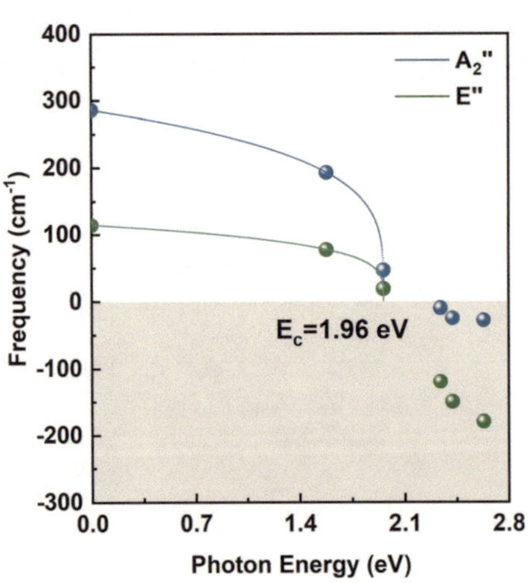

Figure 3.
(a) Phonon frequency of the E" and A_2" modes as a function of photo-excitation energy.

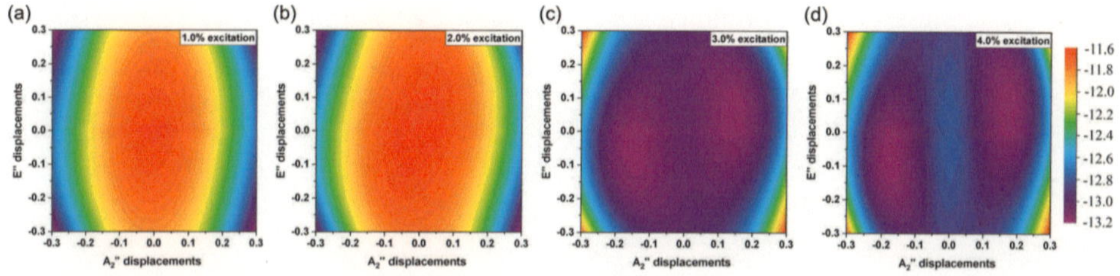

Figure 4.
The two-dimensional potential energy surface spanned by E" and A_2" modes under (a) 1.0%, (b) 2.0%, (c) 3.0%, (d)4.0% photo-excitation per unit cell. Energies are in unit of eV.

The excited electron states that contribute to the electron localization function (ELF) are shown in **Figure 5**. The PIPT is caused because the charge distribution exerts forces to the ions [37]. In **Figure 5(a)** we plot ELF of high-symmetry $MoTe_2$ under different photo-excitation. In the dark environment, the electron gas connects the Mo atoms to form hexagons. As the electron gradually excited from the valance states, the electron gas becomes more localized into the center of Mo atoms triangles. The more electrons excited, the more localized into the triangle vertices. The localization in the triangle vertices is viewed as the microscopic reason that exerts forces on Mo and Te atoms to soften the E" and A_2" phonon modes and induces displacive phase transition.

The second law of thermodynamics shows that free energy is the criterion of spontaneous process, and here we calculate the Helmholz free energy to describe the thermodynamic stability of $MoTe_2$ at 2H and 1T′ phases [38]. As shown in **Figure 6**, at 0 K, the calculated free energy difference between the two phases of $MoTe_2$ is only 33 meV per rectangular unit cell at 0 K, indicating that 2H phase and 1T′ phase of $MoTe_2$ have similar stability at this environment. Furthermore, with the temperature rising, 1T′ $MoTe_2$ becomes more thermodynamically stable than 2H $MoTe_2$ at temperatures higher than 190 K as shown in green line in **Figure 6**. When taking thermal strain into account as shown in orange line in **Figure 6**, the temperature of phase transition decreases to 110 K.

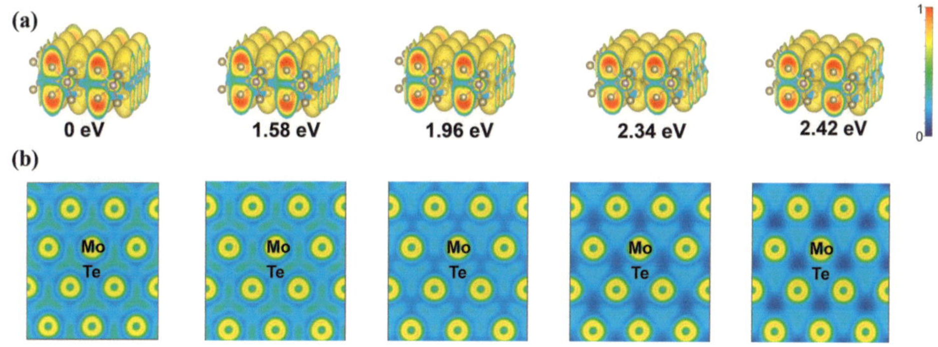

Figure 5.
(a) 3D and (b) top view of the electron localization function of 2H $MoTe_2$ at different excitation energies. The iso-energy surface for the electron localization function in a is 0.2.

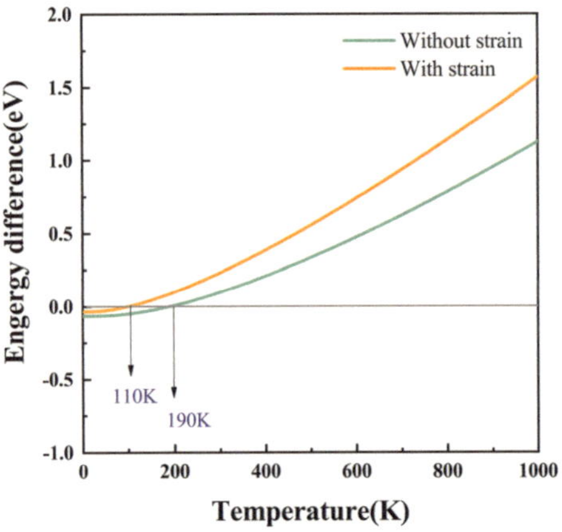

Figure 6.
Helmholtz free energy difference between 2H and 1 T′ $MoTe_2$ monolayers as a function of temperature.

4. Conclusions and perspective

In this chapter, it is demonstrated that a PIPT from 2H to 1T′ phase occurs in monolayer $MoTe_2$ as the excitation energy of photon is higher than the critical value of 1.96 eV by using first-principles calculations. Such structural phase transition is intermediated by the charge distribution transition after photo-excitation, contributing to phonon mode softening and newly emergent potential energy minima. The structural phase transition can be induced under the photo-excitation of $\sim 3.0\%$ electrons per unit cell. Further, this work expresses the phonon frequencies as order parameter and fit it with the external condition indication, namely the photon-excitation energy, and the result fits excellent with the modified Cuire-Weiss law. This research reveals the microscopic nature of the 2H-1T′ PIPT and facilitates the fundamental research of non-equilibrium transient phase transition between normal semiconductor and the topological phase. Also, this theoretical finding is expected to trigger the experimental investigations of PIPT materials in the 2D materials.

By now, stimulated by the advanced experimental methods such as four-dimensional ultrafast electron diffraction (UED), and the recently developed real-time time-dependent functional theory (rt-TDDFT) method, the studies on the PIPT in newly emergent systems such as three-dimensional VO_2 and two-dimensional IrTe have been able to manifest the atomic motions induced by light at femtosecond scale. Furthermore, it has been shown by rt-TDDFT simulations that, by laser pulse excitation, a large part of electrons are excited from valence bands to conduction bands, which not only enlarges the electron occupation in antibonding states, leading to the enhancement of intrinsic Coulumb repulsion and thus enlarged interatomic bond length, but also forms the so-called electron-hole plasma. The dense plasma softens and stabilizes acoustic phonon modes, which drives the distortion of crystal lattice. Recent studies also reveal the competition between coherent light-induced nonthermal collective motions and thermally induced disorded motions, when PIPT takes place. However, there are two important problems still existing. One problem is the nature of PIPT in different kinds of materials. In this work, the microscopic theory based on the Landau phase-transition theory has been developed to explain the PIPT in two-dimensional materials, but whether it is applicable for one- or three-dimensional materials remains unclear. Moreover, the relation between different PIPT-based phases is still needed to clarify. Therefore, it is believed that, it is still challenging to develop a globally microscopic theory for PIPT in this field. The other problem is related to the intentional control over the PIPT-based phases. If the different hidden phases of materials can be induced by light, is it feasible to control by laser pulse to induce a special phase? More experimental and theoretical works are still needed to be conducted.

Author details

Yiming Zhang[1†], Yuanfeng Xu[2*†], Yujie Xia[1], Juan Zhang[1], Hao Zhang[1,3*] and Desheng Fu[4*]

1 Key Laboratory of Micro and Nano Photonic Structures (MOE), Department of Optical Science and Engineering, Fudan University, Shanghai, China

2 School of Science, Shandong Jianzhu University, Jinan, China

3 Yiwu Research Institute of Fudan University, Yiwu City, Zhejiang, China

4 Department of Electronics and Materials Science, Faculty of Engineering, Department of Engineering, Graduate School of Integrated Science and Technology, and Department of Optoelectronics and Nanostructure Science, Graduate School of Science and Technology, Shizuoka University, Hamamatsu, Japan

*Address all correspondence to: xuyuanfeng19@sdjzu.edu.cn; zhangh@fudan.edu.cn and fu.tokusho@shizuoka.ac.jp

† These authors contributed equally.

References

[1] Li W, Qian X, Li J. Phase transitions in 2d materials. Nature Reviews Materials. 2021;**6**:829-846

[2] Krishnamoorthy A, Bassman L, Kalia RK, Nakano A, Shimojo F, Vashishta P. Semiconductor–metal structural phase transformation in $MoTe_2$ monolayers by electronic excitation. Nanoscale. 2018;**10**:2742-2747

[3] Koshihara S, Tokura Y, Mitani T, Saito G, Koda T. Photoinduced valence instability in the organic molecular compound tetrathiafulvalene-p-chloranil (ttf-ca). Physical Review B Condensed Matter. 1990;**42**(10):6853

[4] Nasu K. Photoinduced Phase Transitions. Singapore: World Scientific Publishing Co. Pte. Ltd.; 2004

[5] Ohkoshi SI, Tokoro H, Hashimoto K. Temperature- and photo-induced phase transition in rubidium manganese hexacyanoferrate. Coordination Chemistry Reviews. 2005;**249**(17–18):1830-1840

[6] Tayagaki T, Tanaka K, Yonemura N, Shirai M, Kan'No KI. Symmetry lowering in the photoinduced phase in spin-crossover complexes. International Journal of Modern Physics B. 2001;**15** (28-30):3709-3713

[7] Paillard C, Torun E, Wirtz L, Iguez J, Bellaiche L. Photoinduced phase transitions in ferroelectrics. Physical Review Letters. 2019;**123**(8):087601

[8] Akkanen S-TM, Fernandez HA, Sun Z. Optical modification of 2d materials: Methods and applications. Advanced Materials. 2022;**34**:2110152

[9] Cho S, Kim S, Kim JH. Phase patterning for ohmic homojunction contact in $MoTe_2$. Science. 2015; **346**(6248):625-628

[10] Sie E, Nyby C, Pemmaraju D, Park S, Shen X, Yang J, et al. An ultrafast symmetry switch in a weyl semimetal. Nature. 2019;**565**:61

[11] Yang H, Kim SW, Chhowalla M, Lee YH. Structural and quantum-state phase transitions in van der waals layered materials. Nature Physics. 2017; **13**:931-937

[12] Wall S, Shan Y, Luciana V, Matthieu C, Glownia M, Kozina M, et al. Ultrafast disordering of vanadium dimers in photoexcited VO_2. Science. 2018;**362**(6414):572-576

[13] Liu W-H, Luo J-W, Li S-S, Wang L-W. The seeds and homogeneous nucleation ofphotoinduced nonthermal melting in semiconductorsdue to self-amplified local dynamic instability. Science Advances. 2022;**8**:4430

[14] Kogar A, Zong A, Dolgirev PE, Shen X, Straquadine J, Bie YQ, et al. Light-induced charge density wave in $LaTe_3$. Nature Physics. 2020;**16**:159-163

[15] Zong A, Kogar A, Bie YQ, Rohwer T, Gedik N. Evidence for topological defects in a photoinduced phase transition. Nature Physics. 2019;**15**(1): 27-31

[16] Wu JM, Liou LB. Room temperature photo-induced phase transitions of VO_2 nanodevices. Journal of Materials Chemistry. 2011;**21**:5499-5504

[17] Li G, Xie D, Zhong H, Zhang Z, Xingke F, Zhou Q, et al. Photo-induced non-volatile VO_2 phase transitionfor neuromorphic ultraviolet sensors. Nature Communications. 2022;**13**:1729

[18] Thomsen C, Grahn HT, Maris HJ, Tauc J. Surface generation and detection of phonons by picosecond light pulses. Physical Review B, . Sep 1986;**34**: 4129-4138

[19] Scherbakov AV, van Capel PJ, Akimov AV, Dijkhuis JI, Yakovlev DR, Berstermann T, et al. Chirping of an optical transition by an ultrafast acoustic soliton train in a semiconductor quantum well. Physical Review Letters. 2007;**99**:057420

[20] Vasily V. Temnov. Ultrafast acousto-magneto-plasmonics. Nature Photonics. 2012;**6**:728-736

[21] Gemeiner P, Infante IC, Gusev VE, Ruello P. Giant ultrafast photo-induced shear strain in ferroelectric $BiFeO_3$. Nature Communications. 2014;**5**(4301): 1-7

[22] Rousse CA, Rischel S, Fourmaux I, Uschmann S, Sebbanand G, Grillonand P, et al. Non-thermal melting in semiconductors measured at femtosecond resolution. Nature. 2001; **410**:65-68

[23] Horstmann JG, Böckmann H, Wit B, Kurtz F, Storeck G, Ropers C. Coherent control of a surface structural phase transition. Nature. 2020;**583**:232-236

[24] Liua H-W, Liua W-H, Suoa Z-J, Wang Z, Luo J-W, Lia S-S, et al. Unifying the order and disorder dynamics in photoexcited VO_2. PNAS. 2022;**119**:2122534119

[25] Tan Y, Luo F, Zhu M, Xiaolong X, Ye Y, Li B, et al. Controllable 2h to 1T′ phase transition in few-layer $MoTe_2$. Nanoscale. 2018;**10**:19964-19971

[26] Fukuda T, Kaburauchi R, Saito Y, Makino K, Fons P, Ueno K, et al. Photo-induced tellurium segregation in $MoTe_2$. Physica Status Solidi - Rapid Research Letters. 2022;**16**:2100633

[27] Song S, Keum DH, Cho S, Perello D, Kim Y, Lee YH. Room temperature semiconductor–metal transition of $MoTe_2$ thin films engineered by strain. Nano Letters. 2015;**16**:188-193

[28] Wang Y, Xiao J, Zhu H, Li Y, Alsaid Y, Fong KY, et al. Structural phase transition in monolayer $MoTe_2$ driven by electrostatic doping. Nature. 2017;**550**: 487-491

[29] Kolobov AV, Fons P, Tominaga J. Electronic excitation-induced semiconductor-to-metal transition in monolayer $MoTe_2$. Physical Review B. 2016;**94**(9):094114

[30] Kittel C. Introduction to Solid State Physics. New York: Wiley; 1976

[31] Dove MT. Introduction to lattice dynamics. American Journal of Physics. 1993;**62**(11):1051-1052

[32] Kresse G, FurthmüllerJ. Efficient iterative schemes for ab initio total-energy calculations using a plane-wave basis set. Physical Review B. Oct 1996;54: 11169-11186

[33] John P, Perdew KB, Matthias E. Generalized gradient approximation made simple. Physical Review Letters. 77:3865-3868, Oct 1996

[34] Kan M, Nam HG, Lee YH, Sun Q. Phase stability and raman vibration of the molybdenum ditelluride ($MoTe_2$) monolayer. Physical Chemistry Chemical Physics. 2015;**17**(22): 14866-14871

[35] Yamamoto M, Wang ST, Ni M, Lin Y-F, Li S-L, Aikawa S, et al. Strong enhancement of raman scattering from a

bulk-inactive vibrational mode in few-layer $MoTe_2$. ACS Nano. 2014;**8**(4): 3895-3903

[36] Dove MT. Theory of displacive phase transitions in minerals. American Mineralogist. 1997;**82**(3–4):213-244

[37] Wall S, Wegkamp D, Laura Foglia K, Appavoo JN, Haglund RF, et al. Ultrafast changes in lattice symmetry probed by coherent phonons. Nature Communications. 2012;**3**(1):1-6

[38] van Setten MJ, Uijttewaal MA, de Wijs GA, de Groot RA. Thermodynamic stability of boron: The role of defects and zero point motion. Journal of the American Chemical Society. 2007;**129**: 2458-2465